D1068588

FUELLING
PROGRESS

ONE HUNDRED YEARS OF THE CANADIAN GAS ASSOCIATION

1907  2007

TIM KRYWULAK

FUELLING
PROGRESS

ONE HUNDRED YEARS OF THE CANADIAN GAS ASSOCIATION

1907   2007

TIM KRYWULAK

In 2005, the Canadian Gas Association's board of directors agreed to commission a history in recognition of the association's centennial in 2007. *Fuelling Progress* is not only the history of the CGA, however, it is the story of the natural gas industry within the broader context of Canadian history. To ensure all Canadians have an opportunity to enjoy this book and make use of it as a research tool, the CGA board further agreed to place a complimentary copy at every university, college, and public library in Canada. Thanks to the following companies for undertaking this initiative:

AltaGas Utilities Inc. • **ATCO Gas** • **Enbridge Gas Distribution Inc.** • **Enbridge Gas New Brunswick** • **Gaz Métro** • **Manitoba Hydro** • **Pacific Northern Gas Ltd.** • **SaskEnergy Incorporated** • **Terasen Gas Inc.** • **TransCanada Corporation** • **Union Gas Limited**

FUELLING
PROGRESS

ONE HUNDRED YEARS OF THE CANADIAN GAS ASSOCIATION

1907 100 2007

TIM KRYWULAK

Copyright © 2007 The Canadian Gas Association

Published by
The Canadian Gas Association
350 Sparks St., Suite 809
Ottawa, Ontario K1R 7S8

All rights reserved. No part of this publication may be reproduced in any form without the prior written consent of the publisher. Any requests for photocopying any part of this book should be directed to Access Copyright (the Canadian Copyright Licensing Agency, Toronto).

Library and Archives Canada Cataloguing in Publication
Krywulak, Tim, 1971-
Fuelling progress : one hundred years of the Canadian Gas Association,
1907-2007 / Tim Krywulak.
Includes bibliographical references and index.
ISBN 978-0-9782060-1-7
1. Canadian Gas Association—History.
2. Gas industry—Canada—History .
3. Natural gas—Transportation—Canada—History.
I. Canadian Gas Association
II. Title.
TP700.K78 2007   338.4'76657406071   C2007-900583-7

Personnel
Norman Hillmer: chief consulting editor
Sharon Westbrook: project manager
John Flood: publication manager

Parable Communications Inc.: design
Penumbra Press: picture and copy editing
Douglas Campbell: proofreading

Set in Scala sans, 10.5/14

Previous: Members of the BC Electric Railway Company's
gas department, circa 1927. *Terasen Gas*

Contents

Tables and Sidebars

Acknowledgements

This project would not have been possible without the active support of the Canadian Gas Association, its members, and a great many individuals. The CGA board members, along with president Mike Cleland and vice-president Shahrzad Rahbar, were instrumental in making this book a reality. Norman Hillmer, the chief consulting editor, first put me in touch with them and has done much to refine the project's content along the way. As project manager, Sharon Westbrook made invaluable contributions in co-ordinating the many constituent parts of this work and bringing it all together in a timely manner. John Flood, president of Penumbra Press, provided tremendous guidance through the pitfalls of contemporary publishing. Penumbra's lead editor for this project, Dennis Choquette, did yeoman service in tracking down photos and permissions, bringing new material to light, assisting with the writing of the captions and sidebars, and proofreading and sharpening the text. My research assistants, Mike Ryan, Curtis Gagné, and Sarah Long, also saved me a great deal of time in searching out, photocopying, and filing materials. Carol Grovestine of ATCO, Jane Parry of Union Gas, Joyce Wagenaar of Terasen, and Catherine Newton-Wowryk of TransCanada also graciously assisted in the acquisition of visual materials. Mary Laurenzio, Ryan Mesheau, and the team at Parable Communications saw to an attractive layout and design, and were a pleasure to work with throughout this process. In addition, I would like to thank John Poole, Roland Priddle, Rudy Riedl, Bryan Gormley, Robert Joshi, Wayne Hennigar, Bruce Barnett, George Prociw, and all of those individuals who took time out of their busy schedules to be interviewed and to comment on various parts of this book.

Special thanks to my wife, Cathy Krywulak, along with all of my friends and family, for unwavering support throughout my academic career. The same goes for my PhD and MA thesis advisers, professors Duncan McDowall at Carleton University and James Pitsula at the University of Regina, as well as my former graduate school colleagues Vadim Kukushkin, Jeff Noakes, Darin Peaker, Brian Watson, Rachel Heide, and Emily Arrowsmith, from and with whom I have learned so much. I hope this book makes them proud.

Tim Krywulak

Foreword

Official histories can be controversial. There is something about the word "official" that conjures up hagiographies of sterling leaders and catalogues of unending and unsurpassed accomplishments. Good official histories, however, take their readers and their subjects seriously. They understand nuance. They appreciate the necessity of critical perspective and broad context. They are executed without fear or favour by authors who are rigorously independent. *Fuelling Progress* is official, in the sense that it is the product of an enlightened decision by the Canadian Gas Association to evaluate its past, and it is history as it ought to be written.

The book explains how a small collection of Ontario-based gas manufacturers came to form the foundations of a national association that reflects and represents the imperatives of the Canadian natural gas enterprise. The history of the CGA is a tale of institutional survival, but also of the tumultuous struggle of successive groups of gas industry leaders first to protect and sustain a declining industry and then to establish and build a new one. There were forward-looking policies that led to immediate gains for the association and its industry and, inevitably, there were challenges that arose and objectives that conflicted. By the very nature of the dynamics at play, the way ahead could not be entirely linear or progressive.

As a scholar with a longstanding interest in the place of interest groups in the public-policy process and a firm grasp of the contours of Canadian history, Tim Krywulak grounds his research in the wider milieu, weaving the history of the CGA into the texture of the social, economic, and political thrusts that have transformed Canada over the past century. Shaped by fascinating detail and set against the immense Canadian canvas, *Fuelling Progress* is the compelling history of a private institution that evolved as its country evolved.

Norman Hillmer, Carleton University

FUELLING PROGRESS:
A HISTORY OF THE CANADIAN GAS ASSOCIATION, 1907-2007

I

Foundations

Previous: This view of Toronto, circa 1900, shows what utility companies were fighting for in the early twentieth century: who was going to provide services to the densely inhabited spaces taking shape across the country. *Library and Archives Canada*

1, 2: Caught in the double bind of increasing competition and scarcity of resources, the gaslight industry was forced to re-invent itself. Marketing, often masquerading as health advice, was a key component. *Intercolonial Gas Journal*

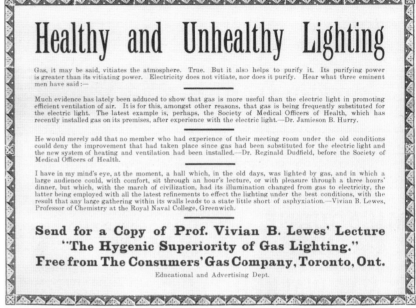

In the early winter of 1907, the editorial pages of Toronto's main newspapers were abuzz with the need to find cheaper power for the city. The *Star* was defending the city's honour against those who said it had "no great interest" in the strictly local plans to develop the hydroelectric potential of Niagara Falls, while the *Globe* was taking the full "blame" for bringing the owners of the Toronto Electric Light Company into conference with the municipal Board of Control to discuss whether or not the city's

AT THE KING EDWARD HOTEL, THE 'GAS MEN' GATHERED TO CONFER ON THE FUTURE OF THEIR INDUSTRY — THEY HAD PLENTY TO DISCUSS

interests could be best served through negotiation or competition. In short, power meant progress for the rapidly growing urban centres of early twentieth-century Canada, and virtually everyone wanted more of both at the best possible price.[1]

At the King Edward Hotel, yet another group of businessmen had gathered to confer on the future of their industry. The "gas men," as they were often known, had plenty to discuss. Most were locally based gas manufacturers who manufactured gas for light, heat, and power from coal. Although their industry had enjoyed remarkable expansion over the last half-century, business conditions were becoming increasingly difficult. Every year, wily inventors, engineers, and scientists were devising new methods for harnessing new forms of energy. Wood, coal, oil, and many other substances were being harnessed in a variety of ways in order to provide the same essential products as those offered by gas manufacturers. Several of the gas-industry leaders gathered in Toronto had seen their street-lighting businesses entirely wiped out by competitors offering an electrical alternative. Now there was talk of creating a giant publicly owned electrical utility in Ontario.

3: The Victoria Gas Co. was the dominant
gas utility in British Columbia when it
was founded in 1860. The gas bill pictured
was presented to Victoria's Masonic Lodge
in March 1863. *CGA archive*

Notice to Consumers. —In case of a leak or injury to the pipes or Meter within the premises of a consumer, the Gas will be
off until the necessary repairs are made.

M *Masonic Lodge*

To the Victoria Gas Company Limited,

No. of Meter _____ Premises *Galls Street*

For Gas consumed from _____ 1863, to *16 March* 18__

Statement of Meter at this date, No. *300*

Less do. at last settlement ....

Consumption .... at $7 50 per thousand cubic feet, $ *2.*

Rent of Meter *3 Mos* —

Bill presented *17/3* 186*3* Total, $ *3*

Received Payment,

No.

☞ **Terms** —Gas will be supplied by Meter at $7 50 per thousand cubic feet. For laying service pipe less than 40 feet, $12;
Meter, $2. Rent of Meter according to size, from 25 cts. to $1 per month.

The Company shall at all times by their Inspector or other authorised agent. have the right of free access into the premises
with Gas for the purpose of examining the whole Gas apparatus, or for the removal of their Meter or service pipe. After the adm
of Gas into the fittings, they must not be disconnected or opened, either for alteration or repairs, without a written permit fr
Superintendent of the Company. Consumers must give immediate notice at the office of the Company of any escape of Gas
deductions will be made from the bills rendered for Gas passing through the Meter. In default of the regular weekly payment
bills, the Company will discontinue the supply of Gas until settlement is made.

▼**Bills presented and collected on the Tuesday of each week.**

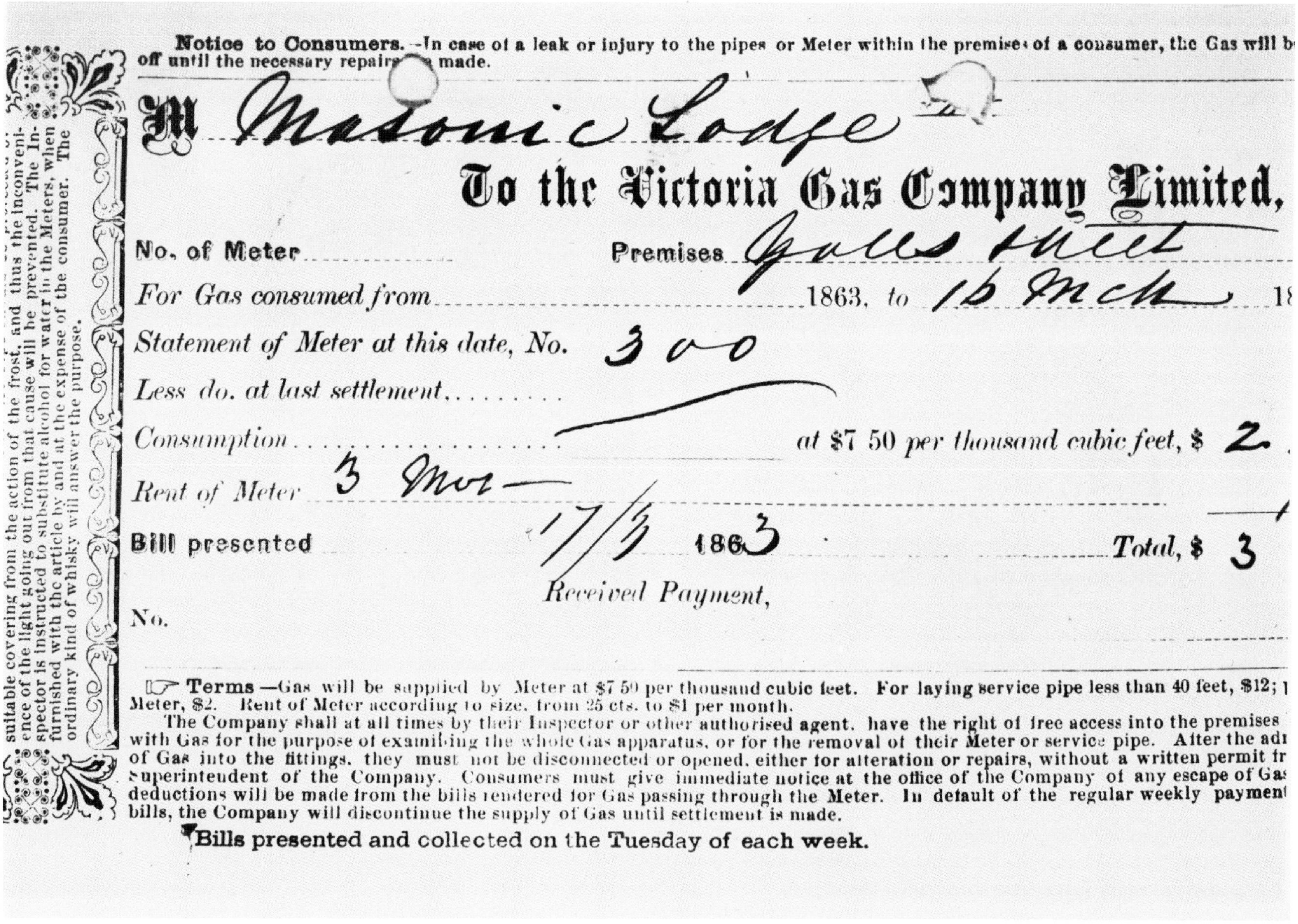

*suitable covering from the action of the frost, and thus the inconveni-
ence of the light going out from that cause will be prevented. The Inspector is instructed to substitute alcohol for water in the Meters, when
furnished with the article by and at the expense of the consumer. The
ordinary kind of whisky will answer the purpose.*

4

4: Immigration played a lead role in the population boom that transformed Canada between 1891 and 1906. This poster, designed to entice Europeans with free farm land in the West, was produced by the Department of the Interior circa 1893. *Library and Archives Canada*

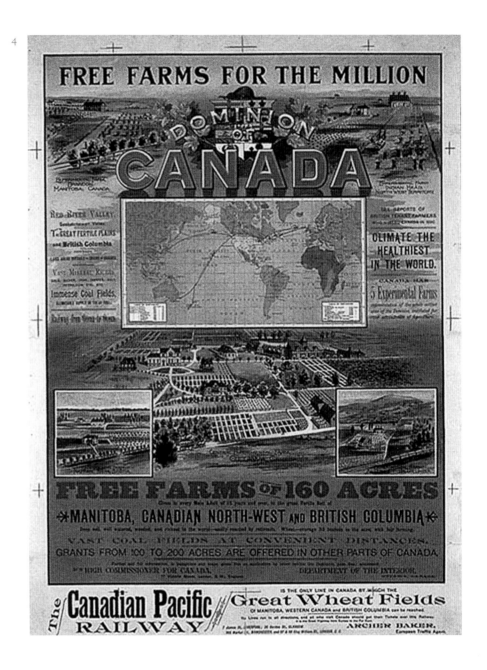

Yet opportunities abounded. Between 1891 and 1906, the population of Canada had grown to more than 6 million people from about 4.8 million – a 25-percent increase. Not only were people having relatively large families (often four or more children), but droves of immigrants (more than 210,000 of whom arrived in 1906 alone) were also flocking to homesteads on the prairies and swelling the workforces of cities, towns, and villages across the country. In the same period, the quantity of wheat produced had grown by 123 percent, the volume of bank notes in circulation had expanded by 114 percent, the miles of railway track in operation had risen by 55 percent, and the gross value of manufactured goods had increased by 53 percent. All of this activity, and the optimism upon which it was founded, was decidedly good for business. Little wonder that Prime Minister Wilfrid Laurier predicted it would be "Canada's century."[2]

The gas-industry leaders who had assembled in Toronto understood the possibilities of their young country, as well as the challenges they faced. Their purpose was to determine how their businesses could continue to contribute to the "virtuous cycle of growth" that had been established, while surviving the always relentless and often bewildering pace of change. The organization they created that day, the Canadian Gas Association (CGA), would form the foundations for pursuing these goals.

Ancient history

The beginning of the gas industry dates back to well before the early twentieth century; it dates back to the formation of life itself. Millions of years ago, the earliest plant and animal life provided the original source of the hydrocarbons that now form the primary ingredient found in coal, oil, and natural gas. As these organisms moved through the cycles of life and death, their remains were gradually covered over in sediment, trapped into various types of sand and rock, and transformed below the earth over a very long period of time. Some were trapped in dense pockets and compressed into hardened form (coal), others were trapped in looser pockets and remained as a liquefied substance (oil), and still others were trapped in open pockets and decayed into gaseous form (natural gas). From then on, each remained there until released by the continued movement of the earth, further transformed by ongoing bio-chemical processes, or discovered and harvested by people.

Human knowledge of the existence and potential uses of hydrocarbon substances dates back to antiquity. In ancient Greece, the eternal flame of the Oracle of Delphi was more likely fuelled by a deposit of natural gas than by any mystical force of the Gods. In medieval Europe, coal provided the heat for fashioning swords and other implements, while oil was used for lighting rooms and for repelling intruders. In tenth-century China, the streets of Peking were illuminated by natural gas conducted via bamboo pipelines. In North America, aboriginal peoples were using oil as an ingredient in paint and for caulking canoes long before contact with Europeans. Such were the beginnings of what would eventually rank among the world's most important industries.[3]

The manufactured-gas industry emerges

The first large-scale applications of manufactured gas occurred in Great Britain, where it began to be widely employed for industrial, heating, and lighting purposes during the late eighteenth and early nineteenth centuries. Manufactured gas offered significant improvements over the motive power of wood and water-driven machinery, the heating power of wood and coal-fired stoves, and the illuminative power of candles and kerosene. Consequently, during the period from the early nineteenth to the early twentieth centuries, its usage spread throughout Europe, North America, and elsewhere.

5: This graphic, included in a CGA education package circa 1979, outlines the geological formation of North America. In the West, most gas reserves are found in sedimentary formations dating from the Cretaceous, Mississippian, and Upper Devonian ages. *CGA archive*

	Time units	Date in years from beginning of period	Succession of life
Cenozoic era	Quarternary	1,000,000	Bat, Man, Ape
	Tertiary	70,000,000	Carnivore
Mesozoic era	Cretaceous	135,000,000	Angiosperm, Horse, Rhinoceros, Frog
	Jurassic	180,000,000	Insectivore
	Triassic	225,000,000	Bony fish
Paleozoic era	Permian	270,000,000	Carnivorous dinosaur, Toothed bird
	Pennsylvanian	310,000,000	Conifer
	Mississippian	350,000,000	Insect, Cotylosaur
	Devonian	400,000,000	Shark, Cordaite
	Silurian	440,000,000	Lung fish, Fern
	Ordovician	500,000,000	Coral, Starfish
	Cambrian	600,000,000	Algae
	Precambrian	3,500,000,000	Trilobite

Manufactured gas was created by burning various organic substances, a process that releases a flammable vapour. This vapour was then captured, stored, transported, and used for the purposes of lighting, heating, and power.

Most of the gas-manufacturing plants in North America were fuelled by coal. Coal-based plants were founded in cities such as Montreal (1841), Toronto (1842), Halifax (1843), and elsewhere from the 1840s to the 1860s.

Oil-based plants were the second-most popular form of gas plant. Initially, their development was constrained by the relatively high cost of raw materials. In the early twentieth century, however, a process developed by New York engineer Walter C. Dayton enabled fuel-oil gas to be safely and efficiently mixed with natural gas. This provided gas companies with a way of "stretching" local stores of natural gas, which supplied the least expensive form of heating available. The Dayton Oil Gas Plant was constructed by the Union Gas Co. for these purposes at Windsor, Ontario, in 1929.

A much more rare plant could be found at Trois Rivières, Quebec. In the early 1900s, the Riché Gas plant, which burned wood, was opened to serve this community. It remained in operation until purchased by a local natural gas company in 1909. It burned down shortly after and was never rebuilt. By the admission of its former manager, L.H. Bacque, the Riché process was "inappropriate" for larger urban centres, but could be useful for smaller communities where local supplies of wood were "abundant and cheap."

Montreal, the commercial and industrial capital of British North America, was the first Canadian city to obtain the benefits of manufactured gas. In the early nineteenth century, the Montreal business community was situated at the very centre of the growing commercial empire of the St. Lawrence Valley. Its ambitions were nothing less than to become the main entrepôt of trade in North America. At this time, Montreal's population was around 40,000, well above those of upstart Canadian rivals Toronto (14,200) and Kingston (6,200), yet well behind that of its main American competitor, New York (312,000). No matter, though, since Montreal's businessmen were still confident they could depend upon British trade policy to provide preferential terms for goods shipped through Canadian ports. So long as this preference – codified in the Navigation Acts – remained in place, they knew it was more economical to ship the produce from large sections of the continent to Europe via Montreal than via New York. In short, they had everything they needed to achieve their goals: the money, the people, and the policy – or so they thought.[4]

By the 1830s, manufactured gas was being enjoyed in London (1807), Paris (1820), New York (1823), and many other great cities of the world. Naturally, progressive-minded citizens in Montreal wondered why their city should not have it as well. As a result, in 1836, the Montreal Gas Light Company was organized to bring gas lighting to the streets of Montreal and into the homes of its wealthier residents. In the customary pattern for that time, the initial charter and capital for the company was provided by a British businessman, Robert Armstrong, while a local resident, Albert Furniss, was appointed as its first manager.

MONTREAL WAS THE FIRST CANADIAN CITY TO OBTAIN THE BENEFITS OF MANUFACTURED GAS

Unfortunately for Armstrong and his fellow investors, not everyone was completely happy with the system of government then prevailing in Upper and Lower Canada (present-day Ontario and Quebec). Farmers complained about the lack of government spending on practical improvements, such as better roads and bridges, while questioning the wisdom of expensive megaprojects, such as the Lachine (1825), Welland (1829), and Rideau (1832) canals. Merchants and professionals protested the British-appointed governor's lack of accountability to the locally elected legislature. Almost everyone objected to the corruption, favouritism, and nepotism that was so obviously rampant among the ruling elites, often known as the "Family Compact" in Upper Canada and the "Château Clique" in Lower Canada. These widely held sentiments ultimately led to the initiation of armed rebellions in both colonies, each of which began in late 1837 and ended shortly thereafter.

Although the Rebellions of 1837 were short-lived, their consequences were far-reaching. In 1841, on the advice of Lord Durham, whom British authorities had dispatched to investigate the troubles in Upper and Lower Canada, the two provinces were renamed Canada West and Canada East, and joined into what was then named the United Province of Canada. The purposes of this measure were to stabilize the economic and political situation by creating a larger tax base and to assimilate French Canadians into the society and culture of English Canada. Instead, it plunged the new colony into a political deadlock lasting for much of the next two decades.

THE POLITICAL UPHEAVALS OF THE DAY MADE IT IMPOSSIBLE FOR THE MONTREAL GAS LIGHT COMPANY TO NEGOTIATE A CONTRACT WITH THE CITY OF MONTREAL UNTIL 1841

A more immediate effect of the political upheavals of the day was to make it impossible for the Montreal Gas Light Company to negotiate a contract with the City of Montreal until 1841. Within one year, however, it soon had more than 300 street lights in operation, which quickly became, as historians Christopher Armstrong and H.V. Nelles note, "a considerable source of civic pride." By this time, Furniss was not only the company's manager but also its principal shareholder, and Montreal's success with the applications of manufactured gas had caught the attention of other cities in Canada.[5]

6: The first 'mechanical chef' was built by storied British gas engineer J. Sharp in 1836. More an oven than a range, the appliance employed a burner much like its twentieth-century counterparts. *Intercolonial Gas Journal*

6

Not a filing cabinet, but the first gas stove

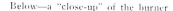

Below—a "close-up" of the burner

7: One of the outcomes of the 1837 rebellions was the creation of the United Province of Canada. More than a political battlefield, however, this region – roughly equivalent to southern Ontario and Quebec – was also the most fertile region for the gas industry in the early days. *Library and Archives Canada*

One such city was Toronto, which one day hoped to displace Montreal as the dominant city in British North America, if not the entire continent. The city, however, lacked both the capital and the expertise to build its own gasworks. It thus sought out Albert Furniss, who was only too happy to assist by organizing the City of Toronto Gas Light and Water Company in 1841. By the summer of 1842, service was initiated.

With two gaslight companies up and running in Montreal and Toronto, and a third in the works for Quebec City, business prospects looked promising for Furniss during the early 1840s. But fate intervened once again. Great Britain, now on its way to becoming the top industrial power in the world, began moving toward freer trade as a way of opening more markets for its products while further driving down production costs. As part of this policy, the entire structure of the Navigation Acts was quickly dismantled over the course of the mid- to late 1840s. Because these acts formed the economic foundation upon which the hopes and dreams of the commercial

8

8: Taking cues from the world's great metropolises – London, Paris, New York – Montreal brought gaslight to the streets in 1836, giving the newly created Montreal Gas Light Co. its key civic mandate. *Library and Archives Canada*

cities of the St. Lawrence Valley ultimately rested, this was no small blow. Moreover, it occurred precisely when nearly 250,000 undernourished and unemployed immigrants were arriving on the shores of British North America seeking refuge from the Irish potato famine. By the end of the 1840s, a sense of gloom and betrayal settled in across the colonies, and city councils and consumers everywhere began to take a much harder look at their expenditures and costs of living.[6]

Many were already annoyed with their gas company. Montrealers demanded to know why their prices were as much as 300 percent higher than those in Great Britain, while Torontonians questioned the quality of service provided by their company. When a second gaslight company, the New City Gas Company, was organized in Montreal by businessman and former city councillor John Mathewson in 1846, Furniss wisely opted to sell his company to his competitor for a reasonable profit rather than to get caught up in a ruinous war of price-cutting and litigation. A similar pattern was repeated in Toronto in 1848, when Furniss sold out to a newly established co-operative venture calling itself the Consumers' Gas Company of Toronto (now Enbridge).[7]

What followed highlights the ironic outcome of many similar efforts in consumer ownership. In its original charter, Consumers' had promised Toronto "gas in greater quantity, of better quality, and at a cheaper rate than the same had been before supplied." True to its word, the company did indeed reduce prices over the first three years of operation. At this time, it had more than 300 shareholders with only 379 customers – making it, in effect, a true "consumers' co-operative." As Armstrong and Nelles explain, however, it was not long before trouble began, as "the number of customers increased beyond the circle of shareholders" and competitors sought to tap into the profits of the local monopoly by threatening to organize yet another rival concern. Consumers' survived, but only at the cost of adopting the practices of any other "profit-maximizing" firm. That is to say it survived by raising prices to a level that enabled it to finance depreciation and expansion, fend off competitors, deal with regulators, and keep shareholders happy. In the meantime, manufactured-gas services spread to Halifax (1843), Quebec City (1849), Kingston (1850), and several other Canadian centres, as did the increasingly fierce debates over prices, regulation, and competition.[8]

9: Charged with lighting Parliament, the Ottawa Gas Light Co. installed 'gasolieres' similar to the ones featured in this advertisement. There were few complaints about the quality of the light – but the heat in the busy centre block was reportedly stifling on occasion. *Intercolonial Gas Journal*

10: Against a backdrop of political and economic turbulence in the 1860s, the idea of a united Canada gained traction. This map shows the disparate elements that would be collected under Confederation a decade later. *Library and Archives Canada*

9

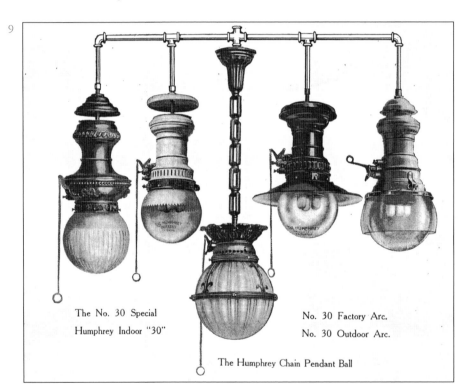

The No. 30 Special Humphrey Indoor "30"

No. 30 Factory Arc.
No. 30 Outdoor Arc.

The Humphrey Chain Pendant Ball

National destinies, national policies

Just as Consumers' Gas was consolidating its position in Toronto, the commercial crisis of the late 1840s and early 1850s was gradually giving way to the relative prosperity of the late 1850s and early 1860s. This economic resurgence was propelled, in part, by the recovery of business conditions in Europe, the expansion of railways at home, and the negotiation of a freer trading arrangement with the United States. By the mid-1860s, however, conditions were changing yet again. The end of the Crimean War in the mid-1850s had weakened the demand for Canadian goods in Europe. Railway investment was proving to be inordinately expensive and notoriously risky. And the United States, angered over what it regarded as British support for the southern states in the US Civil War (1861-1865), sought retaliation by abrogating its trade agreement with Canada. Fears of yet another imminent economic decline returned.

In response, an audacious plan was proposed: uniting the colonies of British North America, building a transcontinental railway, and settling vast tracts of territories then "owned" by the Hudson's Bay Company. According to its main advocates, John A. Macdonald, George-Étienne Cartier, and George Brown, such a plan would not only revive the economic prospects of the colonies but would also put an end to the existing political deadlock by allowing for the "re-separation" of the United Province of Canada into the provinces of Ontario and Quebec, respectively, within the new "Confederation" of Canada.

AN AUDACIOUS PLAN WAS PROPOSED: UNITING THE COLONIES OF BRITISH NORTH AMERICA AND BUILDING A TRANSCONTINENTAL RAILWAY

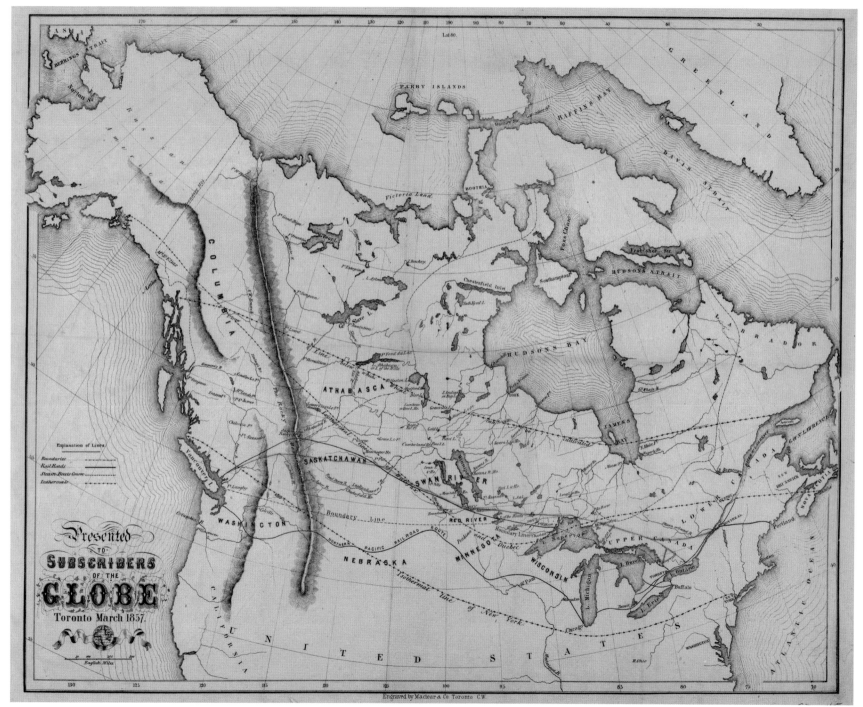

11

11: Napoleon Coste, a chief contractor in the construction of the Suez Canal during the 1870s and 1880s, backed his son Eugene's opening gambit in 1889. With help from his father's capital, Eugene drilled Coste No. 1, which led to the discovery of the Essex gas field. *The Blue Flame of Service*

The Dominion of Canada came into being on 1 July 1867. Its parliament was to be located in Ottawa, the capital of the United Province of Canada since 1857. This was convenient enough, as the construction of a new set of buildings (called "blocks") had been completed to house the province's seat of government only one year before. That said, the centre block, in which the legislature was located, would have to accommodate almost twice the number of representatives for which it was designed. This factor, combined with the gas lighting provided by the Ottawa Gas Company (1854), reportedly made it very hot for those inside.

Before long, the four founding provinces – Ontario, Quebec, New Brunswick, and Nova Scotia – were joined by the Northwest Territories (1869), Manitoba (1870), British Columbia (1871), and Prince Edward Island (1873). By the late 1870s, this territorial expansion was accompanied by Prime Minister John A. Macdonald's "National Policies" of railway

development, western settlement, and tariff protection, bringing along with it two more decades of moderate economic progress. At the same time, there was an ongoing and concerted attempt to identify, classify, and measure the human, financial, and physical resources of the new dominion. As policymakers recognized, if Canada was to be the great industrial nation they hoped it would become, they needed to know what sort of resources it had at its disposal and where such resources could be found. Part of these efforts included the work of the Geological Survey of Canada, which had begun as far back as 1841. It was during this process that the possibilities for the commercial development of natural gas in Canada first came to the attention of a young engineer, Eugene Coste.[9]

The early commercial development of natural gas

The earliest known discovery and practical application of natural gas in North America was recorded by the governor general of Upper Canada, John Graves Simcoe. On 10 November 1794, he wrote:

> among many other natural curiosities, a spring about two miles above the falls attracts the attention of the curious – emitting a gas or inflammable air which, [when] confined in a pipe and flame applied to it, will boil the water of a tea kettle in fifteen minutes.

Similar reservoirs of natural gas were known to exist in New Brunswick, Nova Scotia, and Quebec by the middle of the nineteenth century. One of these reservoirs, located near Trois-Rivières, Quebec, was even used to supply street lighting as early as 1856.[10]

But the natural gas industry did not really get underway in Canada until the late nineteenth century. Its founder, Eugene Coste, was born in Amherstburg, Canada West, in July 1859. Coste was educated in France during the late 1870s and early 1880s, as his father, Napoleon, a French national by birth,

Halifax

By the early 1920s, the Halifax gas industry lagged behind. Reasons abound, but the slump was part of a larger East Coast malaise, firmly entrenched by the 1890s. Before the decline, however, Halifax stood shoulder to shoulder with Canada's major centres, and its gas industry was no exception.

On 10 March 1840, a coalition of prominent Haligonians called for the creation of a gas, light, and water company. Start-up capital of 20,000 pounds was secured a week later, and it was decided that legislature member J.B. Unicake would table a bill to incorporate the utility.

A derelict distillery was purchased a few months later, and, on 9 January 1843, the utility switched on its gasworks. A year later, 281 houses were lighted by gas piped in by the Halifax Gas Light Co., which sold more than 2.8 million cubic feet of gas in its first year.

By Confederation, the company had 2,057 meters, thirty-five miles of mains, and an output of more than 22 million cubic feet of gas.

But prosperity didn't last forever. Competition arrived in 1896, and the company was bought out by the People's Heat and Light Co. shortly after. On the eve of the twentieth century, the overextended endeavour was floundering; it bottomed out in 1902, its assets liquidated to the Halifax Electric Tramways Co.

Gas languished on the back burner for the next decade, until a concerted push in 1916 by the newly established Nova Scotia Tramways and Power Co. once again set the industry on a course for growth.

British Columbia

The early days of the gas industry on the west coast were extremely volatile. As gold and mining fortunes changed – sometimes as quickly as the fates – so too did the prospects for fledgling utilities.

The Victoria Gas Co., founded in 1860 amid Victoria's mid-century placer-mining boom, was the prime gas utility of its day. Early records are scant, but we do know its plant was purchased in Edinburgh, transported by steamer to Victoria, and assembled by an area engineer. The gas it provided was used almost exclusively for lighting, and the utility had to compete with electricity. Whereas standard rates ranged from $2 to $7.50 per 1,000 cubic feet, the Victoria utility offered customers a competitive rate of $1.25 per 1,000 cubic feet.

As the ebb and flow of placer mining descended into a wholesale exodus, the Victoria Gas Co. narrowly escaped bankruptcy on numerous occasions. It scraped by, nevertheless, surviving to see the economy stabilize with the help of forestry and fishing. By 1890, the utility had 1,138 customers and roughly twelve miles of gas mains.

Across the Georgia Strait, meanwhile, the Vancouver-based BC Electric Railway Co. was expanding aggressively, assuming control of gas in Vancouver by 1904, and acquiring the Victoria Gas. Co. in 1905. Although gas was the last phase of the company's expansion, growth came swiftly and substantially. Within five years, the number of customers doubled and main mileage tripled in Vancouver and Victoria, leaving the west-coast gas industry poised to flourish through to the 1920s.

worked as a major contractor on the Suez Canal. After earning a bachelor of science from the Sorbonne and an engineering degree from l'École Nationale Supérieure des Mines, Coste returned to his native country and joined the Geological Survey of Canada in 1883. He remained there for the next six years, progressing through the ranks from geologist to mining engineer. Thereafter, he began a private practice as a mining engineer and developer backed with capital provided by his father, who had also returned to Canada.[11]

Almost immediately, Coste made a stunning breakthrough. Just before leaving the Geological Survey of Canada, he read a report on the geology of Ohio, one that had "sought to establish a relationship between the gas and oil-bearing deposits of that state and a geological structure with a north and south trend known as the Cincinnati anticline." This report prompted Coste to make his initial drilling attempt in 1889, appropriately named Coste No. 1, near his hometown in Essex County. Its outcome was the discovery of what would be known as the Essex gas field. By 1894, there were close to thirty producing wells in the area supplying nearby communities such as Leamington, Kingsville, and Ruthven.[12]

Coste's first natural gas company, the Ontario Natural Gas Company, soon collaborated with several other independent producers to form the United Gas and Oil Company. Their initial hope was to expand into the larger Windsor market, but this endeavour was blocked by determined resistance from the existing gas manufacturer, the Windsor Gas Company, and by the fact that the gas mantles installed around the city did not operate very well with unpurified natural gas. United was forced to look elsewhere for new customers. It found them across the river, in Detroit, in 1894. Within three years, the completion of another pipeline on the American side put United in contact with even more customers. This was good news, until it became painfully apparent that the Essex gas field was not nearly large enough to supply so many customers, especially in light of the relatively inefficient technology of the day. The result was the complete exhaustion of the field by the end of 1904, as well as a ban on the export of natural gas from the province.

The experiences of the United Gas and Oil Company were symptomatic of two of the biggest challenges facing the budding natural gas industry: stiff competition and an uncertain source of supply. Natural gas companies often had to contend with established gas manufacturers and emerging electrical utilities, making it harder to penetrate new markets. The need to rapidly secure such markets was simultaneously reinforced by the recognition that many natural gas companies were sharing the same gas reservoirs. Those who would produce the most profit, therefore, would be those who could draw out the most gas in the shortest time possible. It was a recipe for the kind of destructive competition that benefited consumers in the shortrun, while causing great harm to the industry and its customers over the longer term. But how might one forestall this process?

12: Gas men H.H. Powell, left, and John Keillor were among the first to float the idea of a national gas association, at a meeting held in October 1907. A month later, the Canadian Gas Association was formed. *CGA archive*

12

The Canadian Gas Association is born

The gas manufacturers were, meanwhile, convening to address their own problems. The idea of forming a national industry association to assist them in doing so was first broached during a small meeting convened at the offices of the Hamilton Gas Company in early October 1907. Those present included three prominent gas men: H.H. Powell, the owner of the Brantford and Woodstock gas companies; A.W. Moore, the manager of the Woodstock Gas Company; and John Keillor, the manager of the Hamilton Gas Company. They agreed that the time had come to test the interest in creating a Canadian gas association.[13]

Like their counterparts in other Canadian industries, the gas manufacturers were then grappling with problems of competition and regulation in what was fast becoming a national market. Gone were the days when customers were held hostage by local producers protected by a combination of geography, financial resources, and technological know-how. It was no longer possible for tough-minded gas manufacturers to simply "turn out the lights" when their customers refused to pay an arbitrary 14-percent rate increase, as Furniss had done in Toronto in 1845. The development of the railways, the telegraph, and the banking system, and the widening availability of potential energy substitutes during the last half-century had seen to that, as had the growing willingness of city councils and customers to experiment with alternative forms of ownership and regulation. Now it was becoming more and more difficult just to remain solvent and relevant.

13: In November 1907, the 'gas men' entered the King Edward Hotel. They emerged united under the newly minted banner of the Canadian Gas Association. *Library and Archives Canada*

14: As the gas industry grew, so too did the need for communication among its members. The Gas Journal of Canada was the first serialized publication of the CGA. *Intercolonial Gas Journal*

The Gas Journal
Of Canada
And Waterworks and Sanitary Review
Official Medium Canadian Gas Association

Vol. VII, No. 1.   HAMILTON, JANUARY, 1914   Subscription—$2.00 per year
Single Copies 20c

voice shared concerns, exchange ideas, establish common standards, and protect collective interests.

The CGA was thus officially founded at the King Edward Hotel in Toronto on 21 November 1907. This time there were a dozen gas industry leaders present. Within a matter of a few weeks, the first officers of the association were elected, the association's constitution and bylaws were established, and the first annual convention was scheduled for 26-27 June 1908 in Toronto.

At this convention, inaugural CGA president H.H. Powell declared the purpose of the association was to provide for "the exchange of experiences and suggestions relating to the advancement of the processes of manufacturing and sale of our products and the protection of our interests." According to Powell, the benefits of such efforts were already becoming evident. He noted, for instance, the association's recent work to establish a jointly controlled plant for the processing of gas manufacturing by-products, such as tar, coke, and ammonia, so that gas interests would "no longer be at the mercy of combines who paid next to nothing." He further enumerated the challenges that the industry faced in competing for its heating business in the face of "free anthracite coal, free coke, government bounty-fed oil, and government-exploited waterpower electricity," not to mention the possibility of "free alcohol." The executive committee was thus lobbying for a lower

It is not surprising that the response to this crisis was an organizational one. The establishment of the CGA came at a time when Canadians from many different walks of life were creating myriad political and professional associations. This was because they perceived that an increasingly national political and economic system required national political and economic action. The Canadian Manufacturers' Association (1887), the Canadian Society of Civil Engineers (1887), the Trades and Labour Congress (1902), and the Dominion Association of Chartered Accountants (1902) all emerged around this time. Like the CGA, they provided their members with the opportunity to

15: This medal was given to delegates at the third annual CGA convention. The focal point was a gas exposition, a show-and-tell of the latest developments in the emerging gas industry. *Union Gas archive*

16: As this clipping details, proponents of gaslight in the early nineteenth century were met with a host of objections, some of which seemed irrational by the mid-1920s. *Intercolonial Gas Journal*

15

duty on the gas manufacturers' raw material, bituminous coal, and a higher duty on the residual by-products of gas manufacturing. The latter, as Powell pointed out, was intended to provide for the same opportunities that "other manufacturers and merchants" already enjoyed "under the law." After all, this was at a time when manufacturers of iron, tin, zinc, and other products were bene-fitting from protective duties that often ranged between 20 and 30 percent.[14]

Despite the apparent benefits of collective action, there were also powerful forces working against the formation of national industry, trade, and professional associations. The most significant of these was the "free-rider" dilemma: why bother joining an association, participating in its meetings, and paying its dues if others would do the work instead? To overcome such disincentives, most political associations understood they needed to offer their membership something more than just the distant and uncertain prospect of successful government lobbying. The CGA developed co-operative ventures, such as the by-product processing plant, in order to provide collective economic benefits. It also established an official journal, the *Gas Journal of Canada* (1907-1910), which was later known as the *Intercolonial Gas Journal* (1910-1939) and the *Canadian Gas Journal* (1939-1967), to enable its members to share information and ideas. The CGA's annual conventions allowed busy industry owners and managers to meet in person, relax, and discuss the prominent (and not so prominent) issues of the day. At the first of these meetings, in Toronto, delegates were able to do so while lunching at the Royal Canadian Yacht Club; at the second convention, in Montreal, they did much the same while enjoying a beefsteak dinner served in Dominion Park.[15]

Conspicuous by their absence at these early meetings were most of the recently established natural gas distributors. Aside from a few markets in southwestern Ontario, the majority of gas companies – whether natural or manufactured – were not yet direct competitors with one another. Owing to the still undeveloped state of the existing pipeline infrastructure, along with the inefficiency of trans-mission technology, many companies continued to operate in fairly self-contained markets. This was changing to be sure, but it was a gradual process and one whose outcome remained uncertain. As recently as 1906, the Chatham Gas Company, a gas manufacturer in operation since 1872, had reached an agreement to supply its community with natural gas supplied by the Volcanic Oil and Gas Company, a firm partially owned by Eugene Coste. Only time would tell if conflict or co-opera-tion would mark the future of the industry.[16]

16

GAS AND GOBLINS

Pioneers of Industry Were Met With Strange Objections

More than 100 years ago pioneers of the gas industry who were trying to get the people of Connecticut to discard candles and oil lamps for a new and brighter light were confronted with a paper that set forth the objections to the change in the following manner:

"1—A theological objection: Artificial illumination is an attempt to interfere with the divine plan of the world, which has ordained that it should be dark during the night.

"2—A medical objection: Emanations of illuminating gas are injurious. Lighted streets will incline people to remain out of doors, thus leading to increase of ailments through colds.

"3—A moral objection: The fear of darkness will vanish and drunkenness and depravity increase.

"4—Police objection: Horses will be frightened and thieves emboldened.

"5—Objections from the people: If streets are illumined every night such constant illumination will rob festive occasions of their charm."

It was a long fight, but finally all the goblins of doubt were routed and the gas industry started on its long career of usefulness to man. To-day, although as a lighting medium gas has been largely displaced by a more brilliant illuminant, it is more useful than ever, for it has become America's most popular fuel for heating and has a part in 500 separate and distinct manufacturing operations.

1907

FUELLING PROGRESS:
A HISTORY OF THE CANADIAN GAS ASSOCIATION, 1907-2007

2

Uncertainty

Previous: Taken after striking gas at Bow Island in 1909, this photograph captured an awkward moment. The distance between Eugene Coste, centre, and lead hand W.R. 'Frosty' Martin, far left, was no accident. Martin, a masterly but stubborn driller, kept the operation running full bore despite Coste's orders to shut down. *Glenbow Archives*

1: In the 1880s, railway workers found gas near Langevin, Alta. A nearby landscape is pictured, circa 1910. *Glenbow Archives*

2: A 1912 flare-lighting ceremony, by now a perennial industry custom, marked the arrival of gas in Calgary from Bow Island via a 180-mile pipeline. *ATCO archive*

If Eugene Coste's sole accomplishment had been the first successful commercial development of a natural gas business in southwestern Ontario, that alone would have been an impressive legacy. But there was more. Coste also applied his considerable expertise and good fortune to the search for natural gas in Western Canada. During the early 1880s, workers of the Canadian Pacific Railway (CPR) drilling for water near Langevin, Alberta, found natural gas instead. This led to further explorations, resulting in the initiation of natural gas services to Medicine Hat in 1904. Shortly thereafter, private interests operating in and around Calgary developed yet another reservoir for supplying the city's east end. CPR officials were almost certain there was more gas to be found – gas that could boost their company's revenues through licensing agreements and through fostering the development of the cities, towns, and villages they serviced. In 1908, they hired Coste as their man for finding it.[1]

Notwithstanding his reputation as one of Canada's leading geological authorities, Coste's understanding of the origins of

WITHIN MONTHS OF BEGINNING HIS EXPLORATIONS IN THE WEST, COSTE STRUCK A MAJOR SUPPLY OF GAS AT BOW ISLAND, ALBERTA, IN EARLY 1909

petrochemical substances was based upon a set of assumptions that contemporary geologists now regard as entirely erroneous. As he understood it, the hydrocarbons that formed the molecular basis of coal, oil, and natural gas were generated by chemical reactions between metallic minerals buried within the earth and volcanic substances migrating up from the core of the planet. This was known as the *inorganic* theory of petroleum's origins. Today, however, the great majority of geologists believe in the *organic* origins of petroleum. That is to say, they believe that coal, oil, and natural gas are derived from the interactions

Table 2.1 Gas rates in Canada, 1914: Selected manufactured-gas companies

Name	Location	Net cost ($ per 1000 cu. ft.)		Additional costs	
		Lighting	*Fuel*	*Min. chg. (per mth.)*	*Meter rent (per yr.)*
Barrie Gas Co.	Barrie, Ont.	1.50	1.50	.33	–
Brandon Gas and Power Co.	Brandon, Man.	1.58	1.58	–	–
Belleville Light Department	Belleville, Ont.	1.50	1.25	–	–
Charlottetown Light and Power Ltd.	Charlottetown, P.E.I.	2.16	1.71	–	–
City Gas Co.	London, Ont.	0.90	0.90	–	1.20
Consumers' Gas Co.	Toronto, Ont.	0.70	0.70	–	–
Guelph Light and Heat Co.	Guelph, Ont.	0.85	0.85	0.25	–
Halifax Electric Tramway Co.	Halifax, N.S.	1.70	1.00	0.50	–
Hamilton Gas Light Co.	Hamilton, Ont.	1.00	1.00	–	2.00
Kingston Gas Department	Kingston, Ont.	1.00	1.00	–	–
Montreal Light, Heat, and Power Co.	Montreal, Que.	0.90	0.90	–	2.00
Nanaimo Gas Co.	Nanaimo, B.C.	2.00	1.25	–	–
Napanee Gas Co.	Napanee, Ont.	1.30	1.30	–	–
Ottawa Gas Co.	Ottawa, Ont.	1.10	1.10	–	2.00
Quebec Railway, Light, and Power Co.	Quebec City, Que.	1.20	1.20	0.25	–
St. Hyacinthe Gas and Elec. Light Co.	St. Hyacinthe, Que.	2.40	1.20	–	1.44
St. John Railway Co.	St. John, N.B.	1.75	1.00	–	1.20
Sherbrooke Light, Heat, and Power Co.	Sherbrooke, Que.	1.25	1.25	–	1.50
Vancouver Gas Co.	Vancouver, B.C.	1.75	1.40	–	3.00
Winnipeg Electric Railway Co.	Winnipeg, Man.	1.35	1.20	–	–

Source: Intercolonial Gas Journal

between decaying organic matter and geological and chemical processes that occur over a long period of time.[2]

Be that as it may, Coste's conception of geology certainly qualified as "useful." Within months of beginning his explorations in the West, Coste struck a major supply of gas at Bow Island, Alberta, in early 1909. This time, his achievement was thanks partly to his instincts and partly to the obstinacy of his lead hand, W.R. "Frosty" Martin, who had kept the Bow Island No. 1 well going for a few extra days despite Coste's orders to change locations. Success in business, it seems, has always been a combination of knowing one's industry and hiring the right people.

Over the next two years, Coste secured the leases from the CPR and the capital from investors he needed in order to begin production. With this done, the Canadian Western Natural Gas, Light,

Table 2.2 Gas rates in Canada, 1914: Selected natural gas companies

| Name | Location | Net cost ($ per 1000 cu. ft.) | | | Additional costs |
		Lighting	Fuel	Manufact.	Meter Rent (per yr.)
Beaver Oil and Gas Co.	Brantford, Ont.	0.25	0.25	0.15	–
Canadian Western Nat. Gas, L.H.P. Co.	Calgary, Alta.	0.35	–	0.15	–
Chatham Gas Co.	Chatham, Ont.	0.35	0.25	–	2.40
Dominion Natural Gas Co.	Hamilton, Ont.	0.45	–	–	
Moncton Tramway, Electric and Gas Co.	Moncton, N.B.	0.38	0.15	0.25	–
Medicine Hat Gas Department	Medicine Hat, Alta.	0.15	0.15	0.10	–
Oxford Natural Gas Co.	Woodstock, Ont.	0.25	0.25	–	–
Provincial N.G. and Fuel Co.	Niagara Falls, Ont.	0.40	–	–	–
Sarnia Gas and Electric Co.	Sarnia, Ont.	0.30	0.25	0.125	1.20
Union Natural Gas Co.	Essex, Ont.	0.25	0.25		

Source: Intercolonial Gas Journal

Table 2.3 Electric lighting and power: Cost of electrical services in selected cities, 1913

Locality	Rates ($)
Brandon, Man.	0.10 per kwh
Brantford, Ont.	0.085 per kwh
	less 5% to 35% discount
Fredericton, N.B.	0.12 per kwh (residential)
	0.12–0.16 (commercial)
Guelph, Ont.	0.04 per 100 sq. feet
	plus 0.04 per kwh
	less 20% discount
Hamilton, Ont.	0.085 per kwh (residential)
	0.015 per kwh (commercial)
	less 20% discount
Medicine Hat, Alta.	0.08 per kwh (residential)
	0.06 per kwh (commercial)
Montreal, Que.	0.08 per kwh
	less 20% discount on 5-yr. contracts
	less 5% discount on one-year contracts
Regina, Sask.	0.06–0.07 per kwh by volume (lighting)
	0.035–0.05 per kwh by volume (power)
St. Hyacinthe, Que.	0.15 per kwh
	less 20% discount
Toronto, Ont.	0.04 per 100 square feet
	plus 0.03 per kwh (residential)
	less 10% discount
Winnipeg, Man.	0.035 per kwh
	less 10% to 35% discount
Vancouver, B.C.	0.072–0.088 per kwh by volume
	plus 0.015 meter rental/per month
	less 20% discount

Source: Department of Labour, Board of Inquiry into the Cost of Living in Canada (Ottawa: King's Printer, 1915), 317-26

Heat, and Power Company was incorporated on 19 July 1911. The new company immediately acquired the holdings of two existing franchises: the Calgary Gas Company, a coal-gas manufacturer with approximately 2,200 customers; and the Calgary Natural Gas Company, a supplier with about 50 customers. It also began the construction of a 180-mile pipeline to connect the Bow Island field with customers in Lethbridge (1911), Calgary (1912), and other communities in southern Alberta. The natural gas industry had arrived in the West.[3]

Elsewhere in the country, the development of the gas industry was proceeding apace. By 1914, the CGA's directory of gas companies listed more than seventy-five firms operating across Canada – a considerable increase from the handful that founded the industry in the mid-nineteenth century [Tables 2.1-2.2]. Moreover, although alternative sources of power were in abundance, gas offered a number of advantages over its competitors. Wood, coal, and other combustibles were less expensive than gas for heating and power, but they were also less convenient because they needed to be tended by hand. Oil and gasoline offered convenience similar to gas but were more expensive, because they were already in high demand for other purposes. This left electricity as the gas industry's main competitor.[4]

It was a formidable one at that, frequently offering a roughly equivalent (or sometimes better) combination of convenience, cost-effectiveness, and application, when compared to either natural or manufactured gas [Tables 2.1-2.3]. Throughout the first three decades of the twentieth century, the transmission lines of regional and municipal electrical systems grew inexorably outward from dozens of new public and private generators, threatening the survival of gas companies everywhere. At the same time, the natural gas producers of Ontario remained locked in a desperate

3

3: The gas fields of southwestern Ontario sparked a dramatic showdown between Union Gas and the Southern Ontario Gas Co. in the early 1900s. *Blue Flame of Service*

4: The interchangeable stove was praised for its adaptability. Capable of burning natural gas, coal, wood, and oil, it enabled consumers to match the right fuel to the right use, season, or locale. *Intercolonial Gas Journal*

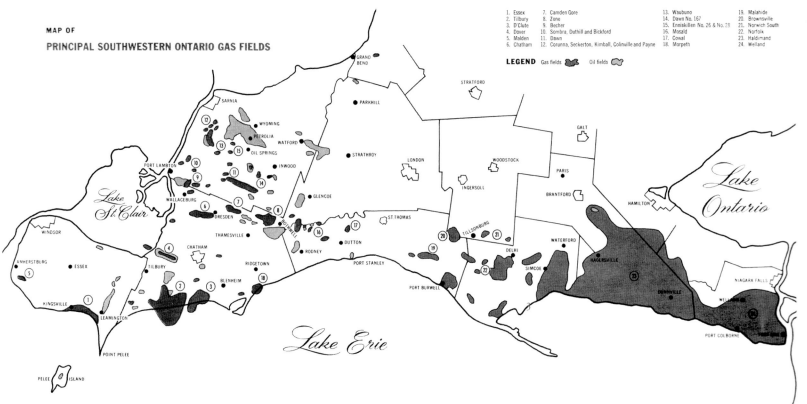

MAP OF

PRINCIPAL SOUTHWESTERN ONTARIO GAS FIELDS

1. Essex	7. Camden Gore	13. Waubuno	19. Malahide
2. Tilbury	8. Zone	14. Dawn No. 167	20. Brownsville
3. D'Clute	9. Becher	15. Enniskillen No. 26 & No. 28	21. Norwich South
4. Dover	10. Sombra, Duthill and Bickford	16. Mosald	22. Norfolk
5. Malden	11. Dawn	17. Cowal	23. Haldimand
6. Chatham	12. Corunna, Seckerton, Kimball, Colinville and Payne	18. Morpeth	24. Welland

LEGEND Gas fields   Oil fields

struggle with one another for control of gas reserves and markets, while gas manufacturers scrambled to finance major investments in technology and marketing just to remain competitive. The demands of the First World War and the postwar boom would only add to these strains. If gas companies failed to meet the growing expectations of their customers, someone else would.

Collaboration and confrontation:
The formation of Union Gas and the Southern Ontario Gas Company

As the competition for the control of the known sources of natural gas in Ontario intensified in the early years of the twentieth century, so too did the search for additional sources of supply. Following the discovery of Coste No. I in Essex County in 1889,

three major gas fields had been opened up across the southern portion of the province: the Welland field, in Welland County (1889); the Haldimand-Norfolk field, in the counties of Elgin and Brant (1890); and the Tilbury field, in Kent County (1905). Of these three, the Tilbury field was to become the largest and most hotly contested.[5]

By the beginning of the twentieth century, numerous firms were involved in the exploration for oil and natural gas reserves in southwestern Ontario. Typical among them was the Acme Oil Company, founded in 1904 by Dr. John Kerr, a Canadian-born dentist working in Detroit, together with several other Michigan-based "gentlemen entrepreneurs." The start-up costs were raised by investors subscribing to a certain number of shares in the company, each of which granted the right to a share in any of its future profits. For those who could afford to participate, such enterprises offered a respectable gamble that might someday offer a handsome return.

WITHIN A YEAR THERE WERE DOZENS OF DRILLING OUTFITS IN OPERATION, AND BY MID-1907 THIS HAD INCREASED TO MORE THAN 200

The Acme company began its drilling operations in Leamington, where several small but promising discoveries had been made earlier that year. Acme enjoyed no such luck. With its capital reserves approaching exhaustion after two unsuccessful attempts, Kerr and his fellow investors were forced to consider their options. Several drinks into a "business meeting" at the Huffman House, the unofficial headquarters of the Leamington-based oil and gas industry, a decision was reached: the $1,000 remaining on the books was not enough to drill a third well – but it was sufficient for one incredible party to celebrate the company's founders' early (but not entirely unexpected) retirement from the oil business. Planning began immediately. The event was to be held at the Ponchatrain, one of Detroit's most luxurious hotels, complete with the very best foods, drinks, and entertainments that the remaining funds could buy. It would be a night to remember.

Sometime between the Huffman meeting and the Ponchatrain affair, Kerr's Scottish, Presbyterian, and pioneer upbringing caught up with him. To spend more money on one night than most workers earned in an entire year would be unconscionable. Why not contribute just enough more money to the venture so that the original investment could be put to productive use in drilling a third well instead? Back at the Huffman, rumour had it that the Essex gas field, the Leamington oil fields, and other nearby wells were all part of the same "petroliferous structures." All one needed to do, then, was tap into these at some point where they could be reached at a reasonable cost. For their last attempt, and to save money on land-leasing costs, Kerr therefore suggested their third well could be drilled on his family's old homestead in east Tilbury. Disappointed, but too shamed to object, the other investors agreed. Fifteen hundred feet down and out of money, they struck oil.

The Acme oil strike of late 1905 soon attracted the attention of scores of "lease-hounds, wildcatters and speculators." Within a year there were roughly 50 drilling outfits in operation, and by mid-1907 this had increased to more than 200. By the end of the year, oil production in the Tilbury field had climbed to 411,588 barrels, surpassing the annual production of the famed "oil springs" at Petrolia, Ontario. Amid all of this excitement, the proprietor of the *Chatham News*, A.C. Woodward, launched an oil and gas weekly, the *Canadian Oil and Gas Derrick*, to service the area's new industry.

Examples of domestic and industrial tasks that can be accomplished with 1000 cubic feet of gas

1. Bake 50 loaves of bread
2. Cook meals for a family of five persons for one week
3. Broil 70 three-pound steaks
4. Boil 275 gallons of water
5. Roast coffee sufficient for 50,000 cups
6. Light two cigars per day for 500 years
7. Bake 500 bricks
8. Anneal 625 pounds of fine wire
9. Hatch two settings of eggs in gas incubator
10. Prepare varnish for 24,000 square feet of flooring

Source: *New York State Committee on Public Utility Information, Intercolonial Gas Journal*

4

Champion Interchangeable

The Gas Consumers have come to recognize its value, for the Champion has more than justified itself by its years of absolute satisfactory service.

It burns Natural Gas, Artificial Gas, Coal and Wood, Coal and Gas together ; a marvellous combination of facilities with one Range.

Its manipulation is simple, for there are no complicated parts to bother the housewife.

Can be changed in three seconds from Gas to Coal, and vice-versa.

It is not merely a "Season" Range, it is ready for service all the year round.

Write our nearest Branch for more particulars.

SEND FOR CATALOG AND PRICE LIST

LONDON, TORONTO, HAMILTON, MONTREAL, ST JOHN, N B   **McClary's**   WINNIPEG, VANCOUVER, CALGARY, SASKATOON, EDMONTON

McClary on Goods is a Quality Name.

Table 2.4 University of Ohio cost comparison for cooking a six-person dinner at US prices in 1917*	
Fuel and Method	**Cost (in cents)**
N/M Gas at 60c per thousand cu. ft. 1 to 2 oz. pressure, proper short flames	$0.0066
N/M Gas at 60c per thousand cu. ft., 4 to 5 oz. pressure, long flame	$0.0132
Electricity at 2c per kw hour	$0.034
Coal at $10.00 per ton	$0.041
Oil at 25c per gallon	$0.090

*Dinner consisting of steak, scalloped potatoes, spinach, rice pudding, and coffee
Notes: In this study by the Department of Home Economics at the University of Ohio, the fuel costs of cooking with natural or manu-factured gas were found to be less than that of electricity, coal, or oil at US prices prevailing in 1917. It is important to note, however, that these figures are exclusive of appliance purchase or rental, metering charges, or any other additional costs. As indicated in a later study by the Ottawa Gas Company, the inclusion of such costs would tend to change the relative pricing structure.

5: The Huffman House was the de facto headquarters of the Leamington oil and gas contingent. It was here that Acme founder Dr. John Kerr and his investors hatched a plan to get out of the drilling racket. After an eleventh-hour crise de conscience, Kerr gambled on another well. It paid off in late 1905, marking the start of the Tilbury oil boom.
Blue Flame of Service

As the title of Woodward's weekly journal indicated, the oil strikes in the Tilbury field were often accompanied by natural gas, which oil producers typically had little use for. Compared to natural gas, oil sold for a much higher price per volume. Furthermore, unlike Tilbury's oil, which could be extracted and marketed with as little or as much additional refinement as its leaseholder saw fit, the commercial development of its natural gas resources would require heavy upfront investments in pipelines, the acquisition of technical knowledge, and the negotiation of complex legal agreements with municipalities, corporations, and residential customers. Therefore, the excess "waste gas"– as it was often called – was simply burned off in giant flares in the belief that this would help to "draw out the oil." By one estimate, nearly 2 billion cubic feet of gas was squandered in this manner during these early years.[6]

For the oil interests, the Tilbury field wasn't the boon it had appeared to be. The Acme strike never recovered the investments of its financiers. Kerr then died suddenly in 1908, just as the overall oil production of the field was heading into a rapid decline. Even the local farmers, who had hoped to profit from the speculations around the best drilling sites, learned the area's "mineral" rights might not belong to them after all. Once it had become clear that these rights had some form of value, a legal battle over their ownership suddenly materialized. A British land settlement company that had colonized large sections of the county now claimed that it was they – not the farmers – who retained the ownership of these rights. The Canada Company, as it was known, maintained that it had ceded only the "surface rights" to the farmers when it had sold them the land; therefore, it retained the rights to any minerals below the topsoil. The farmers, naturally, held the contrary view. The case ultimately made its way all the way to the highest court in the British Empire: the Judicial Committee of the Privy Council. On the issue of mineral rights relating to oil, the court decided in favour of the Canada Company in the matter of *Barnard-Stearns-Roth-Argue Company v. Farquharson* 5 DLR 297. However, it also ruled that because gas was not a mineral, the "gas rights" belonged to the farmers.[7]

6-8: Ambitious gas contractor H.D. Symmes (6) disagreed with D.A. Coste (7) over the direction of Union Gas Co., splintering off to co-create the rival Southern Ontario Gas Co. After the split, A.S. Rogers (8) was appointed Union's first president in 1911.

As it happened, these gas rights proved far more valuable than the oil rights. The Coste interests entered the Tilbury area in late 1906. Within a matter of weeks, their first well, Halliday No. 1, came in with a promising flow of about 6 to 8 million cubic feet of gas per day. By comparison, Coste No. 1, which had supplied customers in Leamington, Kingsville, Ruthven, Detroit, and Toledo for several years, had had an initial flow of around 10 million cubic feet per day. Negotiations with the nearby community of Chatham were opened at once. To strengthen their bid, the Costes simultaneously entered negotiations with H.D. Symmes, an ambitious and "fiery" gas contractor turned operator. In an agreement formalized at the King Edward Hotel, they formed the Volcanic Oil and Gas Company in November 1906. After further consultation with the municipal council, the franchise for Chatham was secured in March 1907.[8]

BETWEEN EARLY 1907 AND LATE 1909, SEVERAL OTHER GAS SUPPLIERS FLOODED INTO THE FIELD, DRILLING WELLS AND EXTENDING PIPELINES IN ALL DIRECTIONS

Between early 1907 and late 1909, several other gas suppliers flooded into the field, drilling wells and extending pipelines in all directions. Volcanic provided service to Chatham, Windsor, and Blenheim. United Fuel Supply captured the franchise for Sarnia; Ridgetown Fuel Supply obtained the contracts for Ridgetown and Highgate; the Maple City Oil and Gas Company supplied the village of Tilbury; and the Beaver Oil Company owned the rights for Leamington and Kingsville. Before long, it looked as though the experience of the Essex gas field was set to repeat itself, much as it was already doing in the Haldimand-Norfolk field.[9]

9: Because rival companies extracted gas from the same reserves, there was an imperative to extract as much as possible, as fast as possible. The Port Alma pumping station was Union's response to a similar project by Southern. *Blue Flame of Service*

9

The greatest threat to the Tilbury gas supply was the Ridgetown Fuel Supply Company's endeavour to link its pipeline to London, Ontario. The plan was for the natural gas to be delivered by Ridgetown and distributed by the current franchise owner, the City Gas Company. Negotiations to this effect began in September 1911. If successful, this would have opened the field to thousands of new customers, almost certainly triggering renewed efforts to secure even larger markets among Ridgetown's numerous competitors. Before making the switch to natural gas, however, the manager of City Gas, J.C. Duffield, sought assurances that the conversion would be worth the costs involved in phasing out the old plant and investing in new infrastructure. Concerned that the Tilbury gas might not last as long as Ridgetown claimed, Duffield demanded an $800,000 bond as a guarantee. Unable to finance both the bond and the pipelines necessary to win the contract, Ridgetown opted to abandon its plans for the time being.[10]

Shortly after the collapse of the London proposal, Ridgetown merged with Volcanic and United to form the Union Gas Company of Canada. The creation of Union Gas made a great deal of business sense: not only did it bring together some of the best minds and most determined operators in the industry, it offered each company the opportunity to pool financial and physical assets. Ideally, these advantages were supposed to enable the new firm to better finance future expansion and to more effectively manage the remaining gas supplies in the field. From the very outset, though, tensions arose over the achievement of Union's two main objectives. Some stakeholders, led by H.D. Symmes, favoured an aggressive campaign of rapid expansion to earn the largest possible return in the shortest possible time; others, led by D.A. Coste, favoured a more conservative approach of accepting a lower return in the short run in the hopes of earning more over the longer term. In the ensuing struggle, Symmes lost. He was not to be part of Union's directorate, and the plan to resurrect Ridgetown's pipeline to London was not to be pursued in the near future.[11]

Yet Symmes was not about to let the matter end there. Using his holdings from Volcanic as collateral, he quickly purchased the other two major interests in the Tilbury field: the Maple City Oil and Gas Company and the Beaver Oil Company. He then looked elsewhere for more capital and additional outlets for Tilbury gas. This search put him in touch with H.L. Doherty, a flamboyant promoter from New York who had built up a sizeable public-utilities empire on both sides of the Canada-US border. By the early 1910s, Doherty's Canadian operations, the Dominion Natural Gas Company and the Glenwood Gas Company, already controlled numerous wells and markets in the Niagara region. For Doherty, the problem was that his main source of supply was in the declining Haldimand-Norfolk region. With Symmes's help, however, his companies could tap into the Tilbury field, which would replenish the supplies for existing markets and make possible further expansion as well.

THE PROVINCE'S TWO BIGGEST COMPETITORS IN THE NATURAL GAS INDUSTRY WERE SET FOR A CLIMACTIC SHOWDOWN — IT WAS UNCLEAR WHETHER EITHER COULD SURVIVE

To this end, the Southern Ontario Gas Company was chartered at Brantford, Ontario, in May 1913, with a capitalization of $13 million. Once organized, it began laying pipe to connect the Dominion and Glenwood systems with Galt and other markets. Similar plans were drawn up to link the Tilbury field with these further-east markets and to extend the entire Doherty system into Hamilton, London, and St. Thomas. More seriously, it also initiated the construction of a new pumping plant designed to extract gas from Tilbury at a higher rate. Union had little choice but to respond, and did so with the construction of its own pumping plant in the summer of 1914. The province's two biggest competitors in the natural gas industry were set for a climactic showdown. It was unclear whether either could survive.[12]

10, 11: Against the backdrop of global war, the Viking News proclaimed Viking, Alta.,the next boom town when drillers struck gas on 29 October 1914. But despite an optimistic and well-attended sod-turning ceremony for the first well, pictured at left, infrastructure lagged, due to skepticism about supply and the increasing expense of doing business in wartime. *Glenbow Archives*

12: The Viking No. 1 well, circa 1915. *Glenbow Archives*

10

11

Doing business in Armageddon:
The gas industry and the First World War, 1914-1918

Across the Atlantic, a much larger struggle was taking shape. For much of the past century, the Great Powers had co-existed without their sporadic disagreements turning into major conflicts. By the early twentieth century, however, the escalating competition for international influence and world markets had made the peace precarious. Simultaneously, Europe had become entangled in a complex system of international linkages in which the "Triple Alliance" of Austria-Hungary, Germany, and Italy now faced off against the "Russo-French Alliance" of Russia and France – with Great Britain officially standing aloof while slowly drifting ever closer to its old rival, France. To complicate matters further, as historians R.R. Palmer and J.R. Colton put it, each power had come to believe that "it must stand by its allies whatever the specific issue ... because all lived in fear of war, of some nameless future war in which allies would be necessary." In the summer and fall of 1914, the immediate issue was the Austro-Hungarian Empire's effort to reassert its influence in Serbia following the assassination of the heir to its throne, Archduke Franz Ferdinand, at the hands of Serbian terrorists. One by one, Austria, Russia, Germany, Belgium, France, and Britain were all pulled into the vortex of war.[13]

McDermid 1252

13: With every part of society contributing to the war effort, Canadian Western service crews followed suit, riding bicycles to conserve fuel. *ATCO archive*

14: Calgary's Palliser Hotel – much like its counterpart in Ontario, the Huffman – was an informal hub for the burgeoning gas industry in the West. *John Bland Canadian Architecture Collection*

13

14

Because it was part of the British Empire, Canada was considered to be at war as soon as Britain had declared war on Germany for its violation of Belgian neutrality in September 1914. As a "self-governing" dominion, however, Canada was free to determine the extent to which it would participate in the war effort. But this was never in doubt. Most Canadians still regarded themselves as British subjects, and thus responded with enthusiasm and determination. When the war began, Canada possessed no air force, a navy of two obsolete light cruisers and two used submarines, and a professional army of only 3,000 soldiers. It would eventually furnish half a million men and women for service overseas, together with billions of dollars in money and supplies – a massive contribution for a nation of fewer than 12 million people.

On the home front, life carried on. Canadians needed goods, services, and employment, and businesses needed income for owners and expenses. Government authorities, moreover, counted on the private sector's assistance in order to mobilize the nation's resources for the unprecedented scale of the war effort. The gas industry was but one example of these realities. Gas ranges were employed in the preparation of food at military bases, and gas furnaces in the manufacture of military equipment. In such cases, the high efficiency and even heat distribution offered by gas made it ideal for applications such as cooking for large numbers of soldiers, melting metal for bullets, tempering steel for bayonets, and annealing copper for shells.[14]

Only weeks after the beginning of the war, a major reservoir of natural gas was discovered in the Viking area, about 120 kilometres southeast of Edmonton. Like their counterparts in many other cities throughout Canada, Edmontonians were anxious to acquire natural gas services as a means of providing affordable power to attract a larger base of industries and residents. For Edmonton, the fact that Calgary, its main rival, had already acquired natural gas made its search more urgent. The city thus entered into an agreement with the Edmonton Industrial Association Drilling Company (EIADC) – a group consisting of the mayor, Billy McNamara, and a group of close to 600 Edmonton residents – to finance the exploration of what its experts had identified as the most promising area. Under the terms of the agreement, the city purchased the lease and EIADC did the drilling. In early November 1914, EIADC's driller, C.M. Flickinger, tore into town with the news that gas had been struck.

Again, things did not go exactly as planned. Since EIADC lacked the funds to develop the well, it proposed to sell its assets to a municipally owned and operated gas utility. But in December 1914 and March 1915, city rate payers rejected this proposal. In the first place, the development of the Viking field was already seen to be somewhat risky and expensive. As was the case in Ridgetown's proposal to supply London, there was concern the gas reserves might not last long enough to make the investment worthwhile. Secondly, the fact the mayor was attempting to sell a company that he partially owned to a public utility of the city he ran did not help matters. McNamara had already been taken to court by ex-mayor William Short over his involvement with EIADC in 1913. And even though McNamara did manage to win the re-election ordered by the Alberta Supreme Court in early 1914, the questions raised by Short's case made it more difficult for EIADC to reach what would be considered a fair deal with the city. With no income being generated by the project, EIADC and Edmonton soon wound up as co-defendants in a mechanics lien brought by the International Supply Company for non-payment of their accounts in late May 1915.

At this point, the Northern Alberta Natural Gas Development Company (NANGDC) assumed the assets and liabilities of EIADC, and obtained the city's rights to the Viking gas field. The principals of the new firm encompassed several prominent backers, including Eugene Coste, Canada's leading gas man, and R.B. Bennett, a future prime minister of Canada. Nevertheless, the legal and political issues relating to the transfer of EIADC's assets and the city's rights took quite some time to resolve. So much so that by the time these issues had finally been settled in early 1917, the costs of pipe, labour, and transportation had risen to the point that construction had become impractical. Recognizing this reality, the NANGDC sold its inventory of pipe while prices were high, determined to hold off on the Viking project until costs came back into line, probably sometime after the end of the war.[15]

Elsewhere in Alberta, an even larger discovery of natural gas had earlier taken place at Turner Valley, approximately 60 kilometres southwest of Calgary. Extensive explorations in this region had begun in the early 1910s, shortly after William Herron found a gas seepage bubbling up from the floor of Sheep Creek. Herron purchased the property without delay, and then joined together with six partners, including experienced driller A.W. Dingman, to form the Calgary Petroleum Products Company. Their first well, Dingman No. 1, came in on 14 May 1914. This well contained a "wet gas" with a high concentration of liquid naphtha, which – once separated from the gas – could be sold as automobile fuel. Hundreds of prospectors rushed into the area, soon making the Palliser Hotel in Calgary the centre of Canada's latest resource boom.

HUNDREDS OF PROSPECTORS RUSHED TO TURNER VALLEY, SOON MAKING THE PALLISER HOTEL IN NEARBY CALGARY THE CENTRE OF CANADA'S LATEST RESOURCE BOOM

While the Herron syndicate made out well over the long run, the field as a whole did not prove immediately productive. As the war progressed, capital for new development became increasingly scarce, and there were no other sizeable discoveries until the early 1920s. Reflecting the growing pessimism in the area, the number of "oil mining" companies in the Calgary City Directory fell from 226 in 1914 to 21 in 1917.[16]

In southern Ontario, meanwhile, the Union Gas Company was attempting to prevent future gas shortages in its market by setting aside a 1,000-acre block of the Tilbury field near the village of Merlin. Believing this portion of the field was unconnected to any other wells, Union hoped to use the area as a reserve to supplement its supplies during periods of peak demand. In April 1916, however, Southern struck gas in the area of Union's reserve, followed shortly thereafter by a second well nearby. When technical problems prevented the normal tubing of this second well, Southern decided to turn its entire volume directly into the company's main pipeline. Once more, Union had little choice but to follow suit, particularly as it became apparent that its proposed "reserve" was in fact part of the larger Tilbury field. With supplies rapidly declining in early 1917, industrial customers were warned that production might have to be either restricted or stopped during the coming winter.[17]

The depletion of the Tilbury field was partially offset for a short time by Union's discovery of the Dover gas field in Kent County. Its first well, Dover No. 1, began producing between 5 million and 5$^{1/2}$ million cubic feet of gas per day in May 1917. Greatly encouraged, Union drilled another dozen wells in the same vicinity over the next several months to little effect. Then its original well suddenly began turning to oil in November 1917, producing between sixty and eighty barrels per day and but two million cubic feet of natural gas. A massive blizzard in the winter of 1918 only worsened the situation.[18]

The provincial government intervened in early 1918, commissioning the Ontario Municipal Railway Board to undertake an investigation of the province's entire natural gas industry. Its report supported the principle of maintaining preference for residential over commercial and industrial customers. Following up on these recommendations in the following month, the legislature passed the Natural Gas Act (1918), which "prohibited the use of natural gas from wells in Kent County, except for residential purposes and for office buildings, stores, schools, hospitals, churches and similar buildings, bakeries, laundries, and dairies." The act further appointed a natural gas commissioner to oversee and enforce the legislation. And although there would be several exceptions, additions, and other revisions in the subsequent versions of the act over the next several years, the principle of residential preference would be maintained. So too would the tendency toward the expansion of the regulatory powers of the state.[19]

By the time the First World War was over in November 1918, more than 60,000 Canadians had died in all branches of overseas service. Many more were left physically and emotionally scarred. Moreover, the war effort itself had placed enormous strains on some of the deepest fault lines in Canadian society. French Canadians abhorred the imposition of conscription, just as English Canadians agreed on its necessity. Farmers demanded greater protection from the "graft and gouging" of bankers, grain handlers, and other "eastern big shots"; workers fought against the erosion of their wartime gains in wages and collective bargaining; and social reformers faced the disappointing reality that the war had not brought about a moral and social transformation of Canadian society. In the federal election of December 1921, large numbers of these voters vented their frustrations at Arthur Meighen's "Liberal-Conservative" government, producing one of the most fractious parliaments in Canadian history. With a thin minority, it would be up to the new prime minister – a former industrial relations consultant and civil servant by the name of William Lyon Mackenzie King – to attempt to put things back together again.

Branching out and coming together:
The CGA and its affiliates in the 1910s

Once largely consisting of Ontario-based gas manufacturers, the CGA gradually became a much broader organization. This process began as early as the 1910s and 1920s, as the organization and its industry became increasingly intertwined with the continental energy markets and regulatory systems that were just beginning to emerge at that time, and would continue to coalesce at an ever-faster pace in the coming years.

Soon after the enactment of the Natural Gas Act of 1918, the natural gas distributors of Ontario decided the time had come for them too to form an association representing the interests of their industry. It was further agreed that it would make sense to join with members of the province's petroleum community as well. A meeting for these purposes was convened at Tecumseh House in London on 18 June 1919. It brought together many of the major figures in the province's natural gas industry, including some old enemies such as D.A. Coste, the president of the Provincial Natural Gas Company and the vice-president of Union Gas; and H.L. Braden, superintendent of the Dominion Natural Gas Company. Also present were T.W. Gibson, the federal deputy minister of lands, forests, and mines; and E.S. Estlin, Ontario's commissioner of natural gas. Oil interests were represented by producers such as J.S. Monroe, the superintendent of the Oil Springs Oil and Gas Company; and James P. Murray, the managing director of the Canadian Oil Producing and Refining Company. Other notables were those such as Frank H. Stover, the "'king' of the drillers"; and E.P. Rowe, billed as "Ontario's most persistent wildcatter." Together, this diverse group established the Natural Gas and Petroleum Association of Canada (NGPA).[20]

The association's first annual meeting took place at the Royal Connaught Hotel in Hamilton on 19-20 September 1919. Its first president, C.E. Steele, the manager of the Sterling Gas Company of Port Colborne, told the assembled delegates that "there are

Table 2.5 Published in the *Intercolonial Gas Journal*, this report details the inspection of gas meters by the Ministry of Inland Revenue in 1914 and reflects the growing regulatory powers of the state in the early twentieth century. Note the remarkably small number of meters rejected by dominion authorities, a reflection of the reliability of the relatively simple "bevel" mechanism that the meters employed to track a customer's gas usage.

STATEMENT showing the number of Gas Meters Presented for Verification, Verified, Rejected and Verified after first Rejection, for the fiscal Year ended March 31, 1914

DISTRICTS	Presented for Verification	Kind		Verified as coming within the error tolerated by law			Rejected			Verified after first Rejection			Totals	
		Wet	Dry	Correct	Fast	Slow	Unsound	Fast	Slow	Correct	Fast	Slow	Verified	Rejected
Belleville	1,892		1,892	703	260	895	3	19	12				1,858	34
Berlin	74		74	12	62								74	
Brockville	80		80	45	13	21	1						79	1
Cobourg	15		15		6	8			1				14	1
Hamilton	9,560		9,560	3,504	598	5,458							9,560	
London	6,828		6,828	1,737	1,133	3,885		36	37				6,755	73
Ottawa	4,236		4,236	267	79	3,890							4,236	
Owen Sound	12		12	12									12	
Peterborough	114		114	44	2	68							114	
Sarnia	39		39	27	3	8	1						39	1
Woodstock	28		28		10	18							28	
Toronto	24,098		24,098	8,202	2,466	13,234		124	72				23,902	196
Montreal	28,067		28,067	4,252	9,395	14,094	111	167	84				27,705	362
Quebec	884		884	363	28	493							884	
Sherbrooke	182		182	75	57	50							182	
St. Hyacinthe	157		157	103	19	35							157	
St. John	1,192		1,192	629	34	528			1				1,191	1
Halifax	209		209	119	27	63							209	
Charlottetown	25		25	2	5	9	7		2				16	9
Winnipeg	4,463		4,463	2,110	933	1,393		12	3	3	4	5	4,436	27
Calgary	569		569	45	41	444		25	1				543	26
New Westminster	4		4	1		3							4	
Vancouver	4,348		4,348	1,007	845	2,389		17	63	7	9	11	4,241	107
Victoria	1,056		1,056	369	273	414							1,056	
Totals	88,132		88,132	23,641	16,253	47,400	123	400	276	10	13	16	87,295	838

Inland Revenue Department,
Ottawa, June 1, 1914.

W. HIMSWORTH,
Deputy Minister.

15

CANADIAN METER COMPANY

LIMITED

Tobey Iron Natural Gas Meter

Regular Artificial Gas Meter

Office and Factory

REPAIRING OF ALL CLASSES OF METERS
ONE OF OUR SPECIALTIES.

METER CONNECTIONS AND FITTING

Agents for METER PROVERS
STATION METERS
WESTCOTT PROPORTIONAL METERS
PRESSURE GAUGES

Tobey "A" Iron Natural or Artificial Gas Meter

Re-inforced Tin Natural Gas Meter

10-Lt. Connections

88-90 CAROLINE ST. NORTH

HAMILTON, ONT.

15: Growth in the gas industry affected many economic sectors, especially manufacturers such as the Canadian Meter Co., which expanded and adapted to meet surging demand for infrastructure. *Intercolonial Gas Journal*

16: G.H. Ferguson, a former energy minister, was Ontario's first natural gas commissioner, a post established by the provincial Natural Gas Act of 1918. *Blue Flame of Service*

17: Bringing friends, competitors, and enemies under one roof, the Natural Gas and Petroleum Association held its first annual meeting in Hamilton at the Royal Connaught Hotel in September 1919. *Blue Flame of Service*

16

18, 19: This elaborate menu card, which includes caricatures of Prime Minister
John A. Macdonald and US President Grover Cleveland, facing page, was
distributed to delegates at an American Gaslight Association meeting held
at Rossin House in Toronto in 1888. In the absence of formal professional
networks in Canada, the rolls of American associations were frequently
populated by Canadians. The CGA and its affiliates were among a tide of
early Canadian associations that sought to fill this national void. *Intercolonial
Gas Journal*

18

many opportunities before us." However, he continued, "the gas men have been fighting themselves in the past years. That is true. We hope that is past." In place of these conflicts, he suggested that the time had come for closer co-operation among "the public and those who serve the public" in the interests of "better management and conservation of the gas business." Many agreed. By this time, the membership of the association had increased to 106 from 68 – a 56-percent increase in three months.[21]

Once founded, the NGPA quickly allied itself with the CGA to collaborate on their frequently overlapping concerns relating to production, conservation, public relations, and legislation. Members of each association were thereafter routinely represented at each other's annual conventions. And, in November 1919, the *Intercolonial Gas Journal* became the "official organ" for both associations.[22]

BY 1919, THE MEMBERSHIP OF THE NATURAL GAS AND PETROLEUM ASSOCIATION OF CANADA HAD INCREASED TO 106 FROM 68 – A 56-PERCENT INCREASE IN THREE MONTHS

A parallel development had recently taken place in the United States, where the National Commercial Gas Association (NCGA), a "commercial association" of gas companies and manufacturers, began to consider joining up with the American Gas Institute (AGI), a "technical association" of gas companies and engineers, in 1916. Much like the Americans who were members of the CGA at the time, Canadians were involved in the NCGA and the AGI as well. Some would come to be prominent figures on both sides of the border. Indeed, it was Arthur Hewitt, general manager of the Consumers' Gas Company of Toronto and a one-time president of the CGA (1910-1912), who offered the resolution for the amalgamation of the two main American organizations. The motion was approved, and the American Gas Association (AGA) came into being in June 1918 with 291 member companies and 936 individual members.[23]

In keeping with what AGA vice-president Charles Holman called the "thought of the times" in favour of "centralization and co-ordination," the AGA became affiliated with the CGA in March 1919. For much the same reason, the CGA had decided to reorganize its membership system in January of that year, for the purposes of attracting more members and encouraging greater participation in its various initiatives. By the end of 1919, the CGA had a membership of 102.[24]

The incremental expansion of the CGA was mirrored by that of the gas industry. In 1907, the *Canada Yearbook* listed 39 gaslight and heating companies employing $10 million in capital and 800 workers, and generating $2.2 million in revenues. By 1919, there were 43 gaslight and heating companies, employing $23 million in capital and 1,138 workers, and generating $10 million in

Every Gas Man In Canada Should Join The N.C.G.A.

The prosperity of the gas business means the prosperity of every individual connected with it. Not only the stockholders, but every officer and employee can take advantage of the growing prosperity of the gas business if he knows how. *Every employee from the lowest to the highest is, in one sense, an investor in the gas company.* To be sure he is putting his time and service into the business instead of cash into stock, but the salaries and wages and promotions all depend upon the same basis as the stockholders, i. e., the expansion and stability of the gas business.

The National Commercial Gas Association is therefore a benefit to every gas man in Canada.

Membership in the National Commercial Gas Association will bring you into contact with the brightest minds in the gas business to-day.

The service of the Association is divided into five parts: (1) General News of the industry, including editorial notes, etc., published in the Bulletin; (2) Technical information; (3) Employment; (4) Education; (5) Convention privileges. All of these are free to members.

Send in your application for membership now.

The cost of individual membership in the Association is $5.00 per annum, payable in advance. Address all applications to

Mr. Arthur Hewitt, Regional Chairman Membership Committee, in Care of

The Consumers' Gas Co. of Toronto
19 Toronto Street, Toronto.

20: Consumers' Gas general manager and one-time CGA president Arthur Hewitt was also a regional chairman of the US National Commercial Gas Association, which merged with the American Gas Institute in 1918 to form the American Gas Association. *Intercolonial Gas Journal*

revenues. On the natural gas side, the total value of natural gas sold in Canada increased to $4.3 million from $1.9 million between 1911 and 1919.[25]

At the CGA's thirteenth annual meeting in August 1920, retiring president V.S. McIntyre could look back with satisfaction and forward with optimism: the war was over, the association was continuing to expand, and the gas industry appeared poised for a new period of growth. The postwar fuel shortage was, however, reason for concern. Since the end of the war, the price of the materials needed for the production of manufactured gas had increased by as much as 400 percent. But thus far, many association members had managed to hold price increases to a bare minimum – absorbing the difference in the hope that "normal" conditions would soon return. In the meantime, the application of more efficient methods of production were sure to help keep costs in line, as were conservation efforts designed to better manage the declining supply of natural gas. The Canadian public's thirst for power was likewise sure to outstrip whatever the nation's hydroelectric companies could provide. That night, the gas men toasted the king, the City of Ottawa, and, of course, the gas industry.[26]

1921

3

Expansion

Previous: Canadian veterans arrive in Halifax via hospital ship at war's end, circa 1918. *Library and Archives Canada*

1, 2: The wartime boom gave way to a bitter postwar recession, creating a need for employment organizations such as the War Veterans' Department and the Employment Service of Canada. *Toronto Star*

WAR VETERANS' DEPARTMENT

BIG SETTLEMENT BOOM, MILLION FOR ONTARIO

West Still Leads, But Ontario Shows Rapid Increase of Loans.

Under the Soldiers' Settlement Act, which provides Government assist-

Veterans or their relatives who are in need of information respecting their pay, allowances, discharge gratuities, pensions, civil re-establishment, vocational training, or any other subject are invited to send their questions to this department. Letters for publication, dealing with issues of special interest to war veterans, provided they are brief, interesting, and fair, are also invited. Address correspondence to War Veterans' Department, Toronto Daily Star, 18 King St. West, Toronto.

A. AND N. VETS TO FRAME POLICIES AT MONTREAL

Favor Unity of Action, to Attain All Veteran Ends.

Many and important are the questions scheduled for discussion and

EMPLOYERS

THE EMPLOYMENT SERVICE OF CANADA

has been created to grade the various classes of workers—trained and untrained—and to place the best in the country at your disposal, through a system of Employment offices from Coast to Coast.

THE PROFESSIONAL AND BUSINESS SECTION

exists to place you in touch with Professional, Business and Technical workers.

THE INFORMATION AND SERVICE BRANCH DEPARTMENT OF SOLDIERS' CIVIL RE-ESTABLISHMENT

has a representative in each of these offices to render whatever special services may be required in the employment of the

RETURNED SOLDIER

Place	Address	Tel. No.
Brantford,	136 Dalhousie St.	2590
Hamilton,	303 Clyde Block	Regent 486
		Regent 1413
"	85 James St. N.	Regent 1877
Niagara Falls,	Newport Building	1221
Orillia,	18 Peter St.	60
Owen Sound,	261 9th St. E.	1125
St. Catharines,	200 St. Paul St.	1269
Toronto,	45 King St. W.	Main 3501
"	287 Queen St. W.	Adelaide 2903
"	845 Lansdowne Ave	Junct. 1087
Welland,	15 Division St.	608

A

The transition to the postwar world was anything but smooth. As the war came to an end, thousands of Canadian workers lost their jobs in war-related industries, just as even larger numbers of soldiers were being discharged from military service. Within months, more than 600,000 workers and veterans found themselves seeking re-employment in an economy that was struggling through one of the worst recessions in recent memory.

All levels of government were ill-equipped to handle this crisis. Prior to the war, policymakers had long assumed that those who truly wanted employment would find it. After all, Canada was a young and developing country in which there was so much work to be done. As for those who were unable to work, social assistance was provided only by family members, private charities, and municipal relief projects. Very few systematic preparations, therefore, had been made for addressing the problem of unemployment on a national scale. Besides that, the federal government had taken on a substantial debt to finance wartime military spending. Public debt charges had ballooned to more than 26 percent of total government expenditures by 1919, compared to only 6 percent in 1913.

Yet some new ameliorative initiatives were launched. A national "employment service" to assist workers in finding new jobs was established in 1918, along with a federal loans program to assist soldiers in resettling on farms. Even though such measures were better than nothing, they could not entirely counteract the disruptive effects the war had wrought upon the world economy. Frustrations grew. Unemployment reached double digits. Those who had jobs faced rising prices, often combined with wage-cut demands made by hard-pressed employers. In the charged atmosphere created by these conditions, more than 3.4 million hours of labour were lost to strikes in 1919, more than any other year since 1901, when the government began keeping records.

By mid-1922, the situation began to improve. International and domestic trade revived, businesses and governments resumed investment, and workers and soldiers re-established themselves in the labour force. At first, the resulting recovery was unevenly spread. The Maritimes continued its long and slow economic decline, while the West showed little improvement until the latter part of the decade. Elsewhere, however, the manufacturing and service industries of Central Canada's burgeoning urban centres expanded at an ever-quickening pace over the mid- to late 1920s. By the end of the decade, this brought the economy as a whole back to a healthy growth rate and reduced the national rate of unemployment to less than half of the immediate postwar levels.[1]

The expansion of public utilities was a big part of the domestic recovery as the impetus to develop Canada's natural resources and "modernize" its cities, towns, and villages resumed when the war ended. During the 1920s, the capacity of the nation's electric stations jumped from 5.5 billion to more than 18 billion kilowatt hours per year – making Canada the second-largest producer of electricity in the world, behind the United States. During the same period, the number of employees at electric stations increased to 16,100 from 10,700, capital employed enlarged to more than $1 billion from $450 million, and total revenues jumped to $122 million from $53 million. Gas services were rapidly extended as well. From 1919 to 1927, gas lighting and heating manufacturers went from 586 workers, $28 million capital, and $11 million in revenues to 3,836 workers, $60 million capital, and $18 million in revenues. Natural gas went from a relatively small $4.2-million industry to one with 1,342 workers, $56 million capital, and $7.6 million in revenues. As the founders of the CGA had predicted during the uncertainties of the early twentieth century, there was room for both electrical and gas producers in Canada's future. A space for the gas industry, however, could only be carved out through intelligent and persistent efforts in the areas of conservation, production, and marketing.[2]

Table 3.1 Number of electrical, manufactured, and natural gas meters in service, 1916-1929

Fiscal Year	Meters		
	Electrical	Manufactured	Natural Gas
1916	505,597	199,514	67,940
1918	661,403	325,244	88,795
1921	860,379	361,479	98,494
1923	1,046,831	379,459	102,007
1925	1,165,664	405,471	106,861
1927	1,314,428	462,496	90,302
1929	1,499,872	504,500	107,504

Source: Canada Yearbook

3, 4: Consumers' Gas advertisements active in the 1910s targeted the highly sought-after industrial market, as well as its emerging – and challenging – residential counterpart. *Intercolonial Gas Journal*

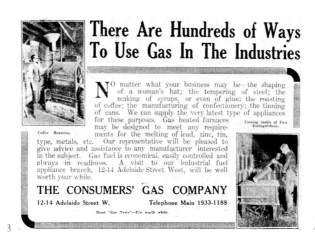

THE EXPANSION OF PUBLIC UTILITIES WAS A BIG PART OF CANADA'S RECOVERY

Save it, sell it, or burn it?
The resource-management dilemmas of the 1920s

As the 1920s began, it was as yet unclear how challenges might be worked out on the natural gas side of the industry. In Ontario the problem was a lack of supply; in Alberta it was a lack of markets. Throughout the next decade, the attempts to address these problems would dominate relations among the industry, the provinces, and the public.

Following the end of the war, the Conservative government of W.H. Hearst moved to return Ontario's natural gas industry to a peacetime basis. The Natural Gas Conservation Act (1918) was replaced with the Natural Gas Act (1919). Under the new

legislation, the pre-war regulatory regime was brought back into effect, with the exception that the wartime restrictions on selling gas to industrial consumers remained.

Gas companies did not like the new act. Industrial consumers were great customers because their demand for gas was high and constant over the course of the entire year. Residential consumers, by contrast, typically purchased the bulk of their gas during the winter months. Worse yet, a large number of residential customers remained tied to long-term municipal franchises that guaranteed service at ridiculously low prices. Such contracts were the legacy of the frantic race for markets that had emerged among natural gas suppliers during the first two decades of the twentieth century. It was one thing to offer low prices when costs were low and supply appeared to be expanding, but quite another to continue to do so as costs were escalating and the supply appeared to be contracting. Nevertheless, with an election looming later that year, the Hearst government was not about to confront urban voters with the prospect of a permanent cancellation of the existing franchises.[3]

When the election was called for October 1919, Hearst and his cabinet were reasonably assured of their chances. In addition to the successful execution of the war effort, the Hearst government had established a solid record of reform legislation. Voting rights had been extended to women for both provincial and municipal elections in 1917, and the Ministry of Labour had been created in 1919. Legislation for mothers' allowances and a minimum wage for women were under preparation as well.

Another term was not to be, however. The sacrifices of the war years had made Ontarians impatient for change. Farmers wanted a greater emphasis on prudent public administration and practical rural improvements, while workers demanded greater assistance and security in the form of more favourable labour and social legislation. To everyone's surprise, and certainly their own, the United Farmers of Ontario (UFO), running in their first election, won forty-five of the province's seats compared to the Liberals' twenty-eight and the Conservatives' twenty-five. Together with the eleven seats won by their allies in the Independent Labour party, this was just enough for a slight UFO-Labour majority.[4]

The "farm-labour" coalition took office in November 1919. Short on political experience, the new government would face a host of complex challenges. Among these were the difficult property rights and contractual issues bound up with the sale and distribution of the province's supplies of natural gas. As the premier, E.C. Drury, saw it, the original ownership of natural gas in his province was vested with the farmers under whose land the resource was located – not the state or anyone else, as was decided by the Judicial Committee of the Privy Council back in 1912.

But once the farmers sold this resource to gas companies, things became much less clear. Ideally, gas companies should receive prices that would allow them to remain in business for the sake of the public interest. At the same time, though, these companies should be held to honour their obligations arising from any contracts they had signed.

The question was how to balance these conflicting rights, interests, and obligations. That winter, gas shortages were reported in Windsor, Chatham, Brantford, Hamilton, and other communities served by the Dominion and Union gas companies. In the spring of 1920, the Drury government – perhaps naively – asked the various parties to work out a negotiated settlement, avoiding the necessity of a legislated solution.[5]

This proved impossible. Hamilton, Sarnia, Petrolia, and a few other municipalities did agree to moderate rate increases; but many other jurisdictions insisted on the continuation of their existing agreements. In response to Union's demands to renegotiate, several of these municipalities formed the Municipal Gas Consumers' Association (MGCA) to resist collectively any changes under all circumstances. Together, they called for a public audit of Union's assets to ascertain whether or not rate increases were actually necessary, as well as an investigation of the desirability of public ownership in the gas industry along the lines of the Ontario Hydro-Electric Power Commission. For its part, Union announced that, with the average test well costing between $12,000 and $15,000 (and sometimes as much as $125,000), it could no longer afford to spend additional money on exploration. To conserve its existing supplies of gas, Union further warned that by 1 August 1920 the company might be forced to disconnect those municipalities that refused to renegotiate.

The gas business was a complex endeavour. Production, competition, government regulations, customer relations, and many other concerns weighed on the minds and moods of owners, managers, and workers. Success called for creativity and a good sense of humour, as evidenced in the pages of the *Intercolonial Gas Journal* of the late 1920s.

IF BILLS WERE ITIMIZED

Total light bill for your home for a month....$5.67

Itemized Statement:

Light consumed in hunting for the dime that
   your small son lost ....$ .34
Light consumed in the parlor on the ten evenings that Jim Perkins called on your daughter, Mary. (Mary doesn't care much for Jim) 2.25
Light consumed in the parlor on the fifteen
   evenings that John Moore called on your
   daughter, Mary. (Mary likes John) ...... .05
Light consumed while you tried to figure out
   an overcharge of fifteen cents on last month's
   light bill .... .18
Light consumed when you forgot to turn off
   the light in the cellar .... .50
Light consumed while you tried to repair leak
   in the water pipes .... .60
Light consumed while plumber (whom you
   were finally forced to call in) told about his
   war experiences and explained how he would
   repair leak 1.20
Light consumed during actual work of repairing leak .... .10
Light consumed while eating, bathing, shaving,
   housecleaning, figuring up household bills
   and accounts, etc. .... .10
Light consumed while spending a nice, quiet
   evening at home with your family .... .05
         —Frank N. Williams, in "Judge."

"THEN AND NOW"

1876

In an old-fashioned house, way down the street,
  Where once had pattered the baby feet,
  Was the gas man's office, grim and gray,
  With a depth of gloom on the brightest day.

Inside so dingy, and dull, and drear,
  That the customer entered in deadly fear;
  So forbidding an aspect, so deathly still,
  For his peace of mind it boded ill.

In the big front room, just off the hall,
  Was a very long counter, with a partition tall,
  With frosted glass and one little hole,
  Where the customer handed over his roll.

The clerk would look up (after a while)
  And glare at the customer with nary a smile;
  "If you don't like the bill, by the Holy St. Peter,
  We'll send down a man and take out the meter."

1916

In the busiest part of the finest street
  Where all the cars of the city meet,
  Is the gas man's home, so gay and bright,
  That it's darker by day than it is at night

You approach the cashier to make a big kick
At the bill, which you think "is just a bit too thick"
  He smilingly offers to reread the meter
  But suggests that you've recently put in a heater

You feel that the gas man is treating you right,
  Giving the best possible value in sight;
That the amount of your bill was not out of the way,
  And the gas man is honest, and not getting gay.

Can you wonder now, in these latter days,
  When competition has so changed our ways,
  That the cuss in customer, and his bag of tricks,
  Were put there by the gas man of seventy-six?

5: As gas companies clashed over dwindling reserves and looming rate hikes in 1920, Chatham – its gas plant shown below – laid claim to supplies from the Tilbury field at prices set in 1907, stoking the fierce debate. *Intercolonial Gas Journal*

5

6: Simmering tensions over gas rates came to a boil on 1 November 1920 when Union cut off its supplies to five municipalities. *Toronto Star*

7: Falling rock pressures in the Tilbury field caused great concern among gas distributors in Ontario. In the mid-1920s, Union Gas recognized part of the problem was due to clogged well sites, and a comprehensive 'blowing out' was undertaken. *Blue Flame of Service*

6

GAS IS SHUT OFF; ONTARIO TOWNS UP AGAINST IT

Tilbury Storm-Centre, and the Townspeople Turn Gas On Again Themselves.

As the deadline came and went, the Drury government continued to urge a co-operative solution. Belle River and Comber agreed to higher rates; Windsor indicated it was prepared to consider the same. But Tilbury, Essex, Blenheim, Ridgetown, and Dresden held firm. Chatham, meanwhile, interjected that its franchise, dating back to 1907, guaranteed preferential access to whatever gas might remain in the Tilbury field, and at the existing rates. To this, the other municipalities rejoined that the wells in existence at the time of that agreement were long since exhausted, making Chatham's legal claims "worthless." Municipal conferences dragged on for the next several weeks with little result.

As another winter approached, Union decided it had heard enough. Asserting its right to cease doing business, the company switched off its gas services to five municipalities on 1 November 1920. To these municipalities, this was a declaration of war. In Blenheim, "special constables" were sworn in to protect the pipelines from company work crews. When the company countered by simply lowering the pressure from its pumping station at Port Alma, a self-appointed "posse" from the town headed over and threatened to "arrest" the control technician. In Essex, an angry mob voiced their displeasure at the offices of the gas manager. Elsewhere, an odd form of hide-and-seek was played out as Union crews severed their own lines as fast as municipal workers could hook them back up again.

Seeking to defuse the situation, the Drury government ordered all gas services restored. On 3 November, Union agreed to comply, without challenging the order in court, upon condition of a slightly higher rate (with the difference deposited in a trust fund) until the entire matter could be settled through arbitration. The government agreed, promising to move forward with new legislation to regulate fair prices and to convene yet another provincial inquiry to examine the conditions in the Tilbury gas field. Days later, it appointed Samuel Wyer, a gas engineering consultant from Columbus, Ohio, to conduct the investigation.[6]

Delivered in May 1921, Wyer's report was a devastating indictment of the wasteful practices found in all branches of Ontario's natural gas industry and among its customers. To begin with, Wyer pointed out that the average natural gas supplier only managed to capture 25 percent of every 1,000 cubic feet of natural gas flowing from each well. Of the gas that made it into the distribution system, local distributing companies lost a further 590 to 4900 cubic feet per year for every three inches of pipe. In these cases, the problem usually stemmed from the fact that the local distributors paid gas suppliers on the basis of a percentage of the gas delivered to the customers' meters rather than for what they received. This meant that these distributors had little incentive to invest in the integrity of their own systems. Consumers, furthermore, wasted an additional 75 percent of the gas that came into their homes as a result of inefficient appliances and poor housing construction. In essence, as Wyer later told the NGPA, just 4 percent of the gas flowing from the province's gas wells was being converted into useable energy.

7

Under current conditions, Wyer calculated that the industry could last "but for a very few years." Therefore, he recommended:

– Gas companies should be banned from open-flow testing, in which wells were blown into the atmosphere for a period of time to determine the average unrestricted flow of gas per day.
– Gas pumped into the transmission system should be measured in several places in order to account for leakage.
– Percentage contracts should be abrogated "in the interests of the public."
– The sale and use of inefficient consumer appliances should be prohibited.
– Prices should be raised to provide a "fair return" on capital invested by gas companies and to encourage conservation.

If adopted, Wyer felt these measures could extend the life of the industry to the early 1940s.[7]

Earlier that year, the government had taken the first steps toward reform with the enactment of the Natural Gas Conservation Act (1921), which had been drafted in collaboration with representatives from the municipalities and the gas industry. It placed the regulation of the industry under the provincial Ministry of Mines, and gave the minister broad powers "to ration gas; to close wells, pipelines, and distribution systems; and to compel installation of appliances for the prevention of waste." It also empowered a "referee" to set natural gas prices and provided that existing contracts could be altered if found to be contrary to the public interest. This method of assigning the regulation of prices to an "impartial expert" followed a pattern that was already well established in other "natural monopolies" such as railways, telephones, and electricity. The hope was that scientific investigation could resolve the disputes over fairness by identifying and justifying a rate that made all parties equally unhappy. According to this standard, referee G.F. Henderson's decision to set Tilbury gas prices between 35 and 50 cents per 1,000 cubic feet could be regarded as a reasonable compromise.[8]

It was soon apparent the supply situation was not quite as bad as originally believed. During the late 1910s and early 1920s, rock pressure tests, which tested the internal well pressure, suggested that the Tilbury region was being rapidly depleted. This was not the case. On closer inspection, Union's superintendent of production, R.L. Bevan, found that many of the wells had simply become blocked off by an accumulation of dirt and various chemical compounds deposited by the gas flow. He ordered the cleaning of those that could be restored and the capping of those that could not. By late 1925, production levels across the field had been substantially revived.

Having addressed its supply problems in the field, Union turned to its transmission system. Many local distributors were still not doing enough to conserve the supply of gas. Union therefore undertook to acquire those companies so that it could upgrade the entire network from wellhead to customer, as well as embark on a more aggressive expansion into the markets controlled by local franchise holders. Between 1919 and 1923, it absorbed the Tilbury Town Gas Company and the Northern Pipeline Company for these purposes. From 1925 to 1927, it purchased the Sarnia, Windsor, Wallaceburg, and Chatham gas companies. These moves would help to secure the company's position as a dominant player in Ontario's gas industry.[9]

On the other side of the country, a much different set of circumstances prevailed. In Edmonton, the NANGDC had plenty of gas. What it did not have was the $4 million it needed in order to construct a pipeline from its field near Viking, roughly 110 kilometres distant. This problem was resolved in 1923, when the NANGDC was bought out by Northwestern Utilities, a subsidiary of US-based International Utilities.

Formed in May and opening its head office in June, Northwestern committed to delivering gas to Edmonton consumers by that winter. This astonishingly difficult task was assigned to the company's first president, E.G. Hill, of the New York-based consulting firm Ford, Francis, and Bacon. Construction began in early July, with careful timetables drawn up to ensure a continuous work process based on the flow of pipe from Canadian, American, and Scottish mills. Transportation delays and poor weather placed the project in jeopardy on a number of occasions, but did not prevent its completion by late October. A system consisting of 180 miles of pipe, fifteen regulator stations, a warehouse, workshop, and other buildings had been erected in just eighty-eight days. Edmonton's Mayor, D.M. Duggan, switched the gas on in an official ceremony on 9 November – almost one decade after viable supplies had been first discovered by city residents.

 8

 9

The management of Northwestern was then turned over to C.J. Yorath, a most unlikely successor. Yorath was an outspoken advocate of public ownership in utilities. In fact, as Edmonton's commissioner of public utilities, he had spent much of the preceding year and a half doing his best to prevent Northwestern from being awarded the franchise for providing natural gas to the city. In September 1924, however, he was convinced by H.R. Milner, Northwestern's legal counsel and future president, that his concern for the public interest was precisely what qualified him for the job. Under Yorath's direction from 1923 to 1932, the company expanded from 1,880 to more than 10,000 customers and from 51 million to more than 3 billion cubic feet of annual gas sales.[10]

In Turner Valley, there was also an abundance of gas to be found. The Herron syndicate's Calgary Petroleum Products (CPP) company had built a small gas-processing facility in the mid-1910s to extract naphtha from its Dingman No. 1 and No. 2 wells. Despite its modest success, outside investors remained skeptical. When the plant burned down in 1920, CPP thus had few options for raising additional capital. Imperial Oil was the only company to show even the remotest interest in the venture. Eventually the owners of CPP agreed to trade their company for one-fifth of the shares in the Royalite Oil Company, owned by Imperial, in order to finance a new plant and further explorations.

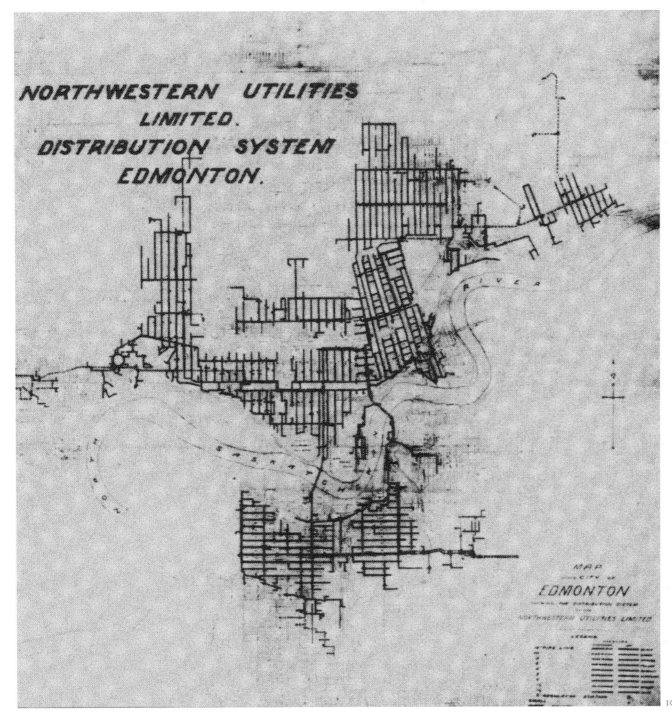

NORTHWESTERN UTILITIES LIMITED. DISTRIBUTION SYSTEM EDMONTON.

8: C.J. Yorath, a staunch advocate of publicly owned utilities, headed the privately owned Northwestern Utilities from 1923 to 1932. *The Courier*

9: Northwestern went from having 1,880 customers in the Edmonton area in 1923 to more than 10,000 less than a decade later. Logo circa 1940. *The Courier*

10: This map shows each district of Northwestern's coverage in Edmonton by the late 1930s. *The Courier*

11: The equivalent of eighty-seven boxcars' worth of coal was flared off each day at Turner Valley, giving the fiery, otherworldly site the nickname 'Hell's half-acre.' *Glenbow Archives*

12: In the early days, a lone Northwestern Utilities truck comprised the entire residential service fleet (circa 1924). *ATCO archive*

11

It was a good decision. On 14 October 1924, a large volume of gas blew in at Royalite's No. 4 well. The pressure was so great that workers had difficulty in controlling it at first. As soon as they clamped the well, the pressure gauge began to rise rapidly. Minutes later, "the drillers ran for their lives," as eighty-five tonnes of pipe was blown out of the ground and came crashing down around the well. Then the gas caught fire. A special crew brought in from Oklahoma would need more than two months to contain the well. Once they did, the gas flow was measured at 20 million cubic feet per day. By the end of the decade, there would be close to one hundred producing wells in the area and the CPP investors' 20-percent share in Royalite would be worth more than $12 million.[11]

GIANT FLARES COULD BE SEEN FROM MILES AWAY. FARMERS COULD HUNT BY NIGHT WITHOUT ADDITIONAL LIGHT

Much as in earlier years, the prospectors of Turner Valley found it more profitable to sell liquid naphtha than natural gas. The naphtha was extracted; the gas was burned. These giant flares could be seen from miles away. In the summer, farmers could hunt by night without need of additional light; in the winter, the grass nearby remained green and uncovered by snow. According to one expert, petroleum engineer Stanley J. Davies, the amount of gas burned each day in the valley during the late 1920s was equivalent to 87 1/2 boxcars of coal. The spectacular view created by this process became of a tourist attraction, earning the area the name "Hell's half-acre."[12]

A revolution in exploration and extraction: The introduction of rotary drilling

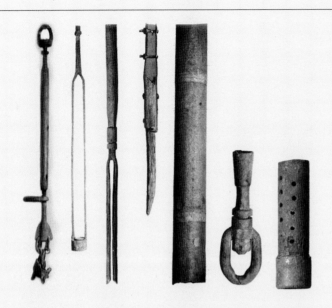

The basic technology for drilling oil and gas dates back more than 2,000 years, when the Chinese used cable tools to drill for salt water. A sharpened tool was attached to a wooden tripod, raised, and then dropped to the ground, pulverizing and digging into the earth below. As the tool sunk deeper, rocks and fine sand were cleared away with the help of a "sand trap," or by pouring water into the hole and scooping out the muddy residue with a bailer. Sections of bamboo, later pipe, were then inserted to keep the hole from collapsing. The entire process was repeated until the desired depth was achieved.

From the mid-nineteenth to the early twentieth century, the same methods were employed by many Canadian oil and gas pioneers. Up to that point, the main innovations had been replacing the wooden tripod with a large wooden derrick and using steam power to raise the cable tools. At the turn of the century, inventors were seeking to improve these methods by experimenting with hollow-tipped drill bits. The "twin-cone" roller bit, patented by Howard Hughes Sr. in 1909, was among the most successful.

When coupled with steam power, the rotary drill provided a major breakthrough. Though requiring a larger investment, the automated rigs could be operated non-stop and at a much faster pace than traditional cable rigs. Rotary drilling completely replaced the use of cable tools by the late 1940s, when the last cable-tool well was reportedly sunk in the mountains of southwestern Alberta. Examples of cable tools are pictured above.

Sources: John Schmidt, Growing Up in the Oil Patch (Toronto: Natural Heritage/Natural History, 1989), 31-3; Sandy Gow, "The History of Drilling Technology," Innovation Alberta, 14 Dec 2006; and Kris Wells, "Making the Hole Was Hard Work," E&P Global Exploration and Production News, 14 Dec 2006.

Several attempts were made to identify more productive uses for Turner Valley's gas. One set of plans called for piping it to Vancouver and Winnipeg. Another called for piping it to Montana and other US states. And still another called for the encouragement of "a large group of British capitalists," who were said to be interested in using the gas "to engage in certain chemical manufacturing processes." Yet, because of the small markets, technological challenges, and geographical distances involved, little came of these proposals, except for the piping of a small amount of gas to Montana to meet peak demand – even as gas suppliers and consumers elsewhere in Canada and the United States strove to conserve local supplies and to search for cheaper sources of power.[13]

Nearby, in Calgary, the supply situation had become troublesome for the Canadian Western company during the late 1910s and early 1920s. River floods washed out several pipelines in 1917, disrupting service and causing a large expense. Then the pressure levels in the Bow Island field began declining in 1920. As in Ontario, supplies were thereafter restricted largely to residential consumers to conserve the remaining gas.

Faced with rising costs, falling revenues, and the depletion of its main source of supply, the company sought a rate increase from approximately 35 cents to 65 cents per 1,000 cubic feet to finance new exploration. The city responded by convening a technical conference in May 1921 to consider the matter, with a committee that included two representatives of the city, two from the company, and an independent chairman. When the conference reported back with a unanimous recommendation in favour of a rate increase, however, the city still rejected the proposal. Only at a subsequent hearing before the board of public-utility commissioners in September did the city reluctantly agree. In return for an increased rate of 48 cents, Canadian Western was to construct a new line to connect the Calgary distribution network to suppliers in Turner Valley by no later than December of that year and undertake "energetic" explorations in the Chin Coulee and Foremost gas fields. Should the company fail to meet either objective, the rate would immediately revert to 35 cents.

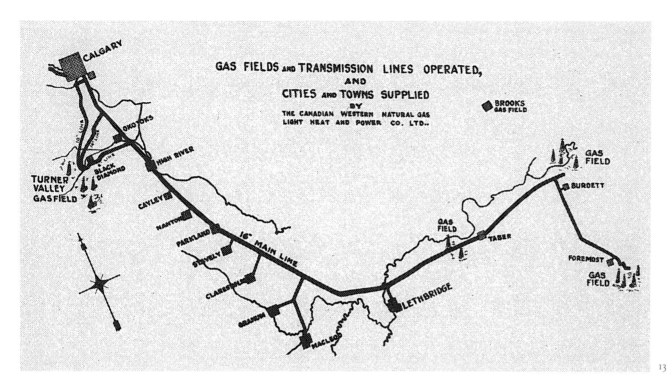

GAS FIELDS AND TRANSMISSION LINES OPERATED,
AND
CITIES AND TOWNS SUPPLIED
BY
THE CANADIAN WESTERN NATURAL GAS
LIGHT HEAT AND POWER CO. LTD..

13, 14: Faced with dwindling supplies and falling revenues by the 1920s, Canadian Western sought a rate increase in Calgary. The increase was tied to aggressive expansion and exploration, fuelling an expanded network of supply and distribution. Map and logo circa 1940. *The Courier*

13

14

Work on the pipeline began in earnest but was frequently delayed by bad weather and a lack of trained labour. A crew of pipe layers was brought in from Winnipeg in an effort to speed production. This was to no avail, as it turned out that these workers "knew more about laying water lines than gas lines." Company crews redoubled their efforts, finishing the line late in the afternoon on the last day of grace.[14]

At this juncture, Eugene Coste decided to step down as the president of the company he founded in 1912. By this time, he was a wealthy man with vast success as a geologist and as an entrepreneur. Now he sought to be closer to family in Ontario, and to spend more time travelling and enjoying life. The sudden death of his son, Dillon, in 1920, possibly hastened this decision. Upon his departure from Calgary in 1922, Coste offered his twenty-eight-room mansion at 2208 Amherst Street for use as a children's hospital. The city declined, but Coste's longtime chauffeur did accept his offer to live in the estate's coach house.

He continued to do so until the property was seized for non-payment of taxes in 1935, which ended the Coste era in the West.[15]

Canadian Western's battles with the city over pricing and supplies persisted. Although the company's drilling efforts at Chin Coulee produced no positive results, those at Foremost generated a number of "splendid" wells. Together with the supplies purchased from Royalite's wells in Turner Valley, the company was able to maintain its services and avoid the termination of its franchise.

In 1925, International Utilities acquired a controlling interest in Canadian Western. C.J. Yorath replaced H.B. Pearson as president, and the new ownership injected much-needed capital and expertise. The company began to expand its pipeline capacity from Turner Valley to Calgary. With the increase in supply, gas rates were lowered to 43 cents. Toward the end of the decade, these rates were further reduced to 33 cents, as annual revenues rose to more than $2 million.[16]

British Columbia Electric Railway Co Ltd

Miles of
Gas Mains

Bar chart showing Miles of Gas Mains by year:

- 1906: 52 Miles
- 1907: 58.2
- 1908: 70.2
- 1909: 86.5
- 1910: 128
- 1911: 158.9
- 1912: 178.1
- 1913: 187.4
- 1914: 190
- 1915: 192.7
- 1916: 192.9
- 1917: 192.9
- 1918: 193.7
- 1919: 194.4
- 1920: 197.5
- 1921: 204.9
- 1922: 204.9
- 1923: 208.4
- 1924: 234.1
- 1925: 256.1

YEARS

(vertical axis: Miles of Gas Mains)

15: By the mid- to late 1920s, gas companies were facing significant capital costs. More pipe meant higher costs, as this graph depicts, forcing companies to increase revenues through innovation and marketing. *Intercolonial Gas Journal*

16: Eugene Coste offered his sprawling Calgary estate for use as a children's hospital when he left for Ontario in 1922. The city declined the offer. The mansion was seized in 1935 for unpaid taxes. *Glenbow Archives*

THE PROPERTY WAS SEIZED FOR NON-PAYMENT OF TAXES IN 1935, WHICH ENDED THE COSTE ERA IN THE WEST

Innovation and amplification:
Building the gas market in the 1920s

As their transmission networks were extended and their capital requirements for depreciation, exploration, and other aspects of their business were enlarged, gas companies had to find ways to increase revenues as well. Many did so by investing in improved technologies to lower operating costs and provide better service; others turned to more active marketing to increase sales.

Because gas tended to be more expensive than its possible substitutes, the need to reduce operating costs was particularly acute among manufacturers. To remain competitive, many chose to invest in new plants and equipment. This was a high-stakes strategy. If it failed, the substantial costs involved in the attempt could ruin the company; if it succeeded, the efficiency gains it provided could secure the company's future for another ten to twenty years. That said, for those possessing aging plants that had reached the limits of their capacity, efficiency, and longevity, the status quo was an option few could afford.

17

17, 18: While the Ottawa City Gas Co.'s ambitious new plant, bottom, was designed to drastically outperform the original, top, its hefty price tag made it a tremendous risk. *Intercolonial Gas Journal*

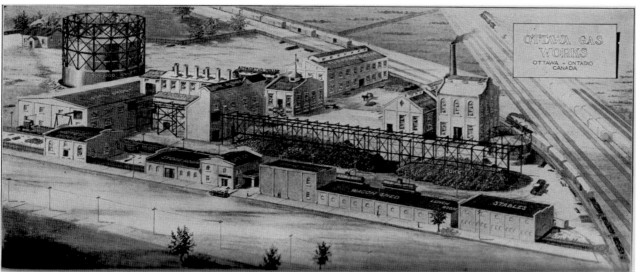

18

The Ottawa City Gas Company, one of the oldest gas companies in the country, was one example. During the mid-1910s, it was still carrying on business with a plant, manufacturing process, and equipment that dated back to the early 1850s. Supplying gas for residential cooking and water heating was the company's main source of revenue, though it did have some industrial and commercial customers, such as the Royal Canadian Mint and the Ottawa Car Manufacturing Company.

So long as there was little competition, this business was more or less secure. In the late nineteenth century, however, the situation began to change. In 1894, the Ottawa Electric Company entered the market and soon captured the municipal lighting contract. Then, in 1906, Ottawa Gas and Ottawa Electric were merged into the Ottawa Light, Heat and Power (OLHP) Company, once again placing a single entity in control of these utilities. A few years later, the city decided to become a local distributor of power as well, as a means to ensure that low prices were maintained. This move prompted the OLHP to undertake more assertive efforts to maintain its position, including the upgrading of the Ottawa City Gas Company's plant.

Driving innovation, embracing efficiency: The transition to vertical gas plants

By automating the process of coal handling, vertical gas plants greatly improved the efficiency of gas manufacturing. This schematic shows Canada's first "self-clinkering" plant, purchased by the British Columbia Electric Power and Gas Co. and erected at Vancouver in 1928. The plant was divided into two units, each with an output of 1,900 cubic feet of gas per day.

1. BELT CONVEYOR
2. COKE HOPPER
3. COKE CHUTE
4. COKE SCREEN
5. BREEZE CHUTE
6. HYDRAULIC COKE CHARGER
7. CHARGER CONE VALVE
8. GENERATOR
9. AUTOMATIC OPERATOR
10. ANNULAR BOILER
11. GRATE BARS
12. GRATE DRUM
13. INNER SEAL
14. OUTER SEAL
15. BLAST STEAM & GAS CONDUIT
16. CARBURETTER
17. OIL SPRAY REGULATOR
18. SUPERHEATER STACK VALVE
19. BOILER STACK VALVE
20. GAS OFF TAKE
21. BACK RUN MAIN
22. BUTTERFLY VALVE
23. WASHER
24. GAS OUTLET TO CONDENSERS
25. BLAST TO GENERATOR
26. BLOWING PLANT
27. WASTE HEAT BOILER
28. CENTRIFLOVANE GRIT CATCHER
29. TO DUST RECEIVER
30. STEAM DRUM

Source: Intercolonial Gas Journal

While the old "horizontal" coal plant employed dated production methods and required large amounts of labour, the new "vertical" plant employed a modern manufacturing process that needed far less labour. This transition was overseen by the West Gas Improvement Company, an engineering consulting firm from Manchester, England. Upon completion in 1919, the plant delivered a 33-percent reduction in coal requirements together with a 23-percent increase in output. An equally promising outcome was realized in Victoria, British Columbia, in 1921, when the Victoria gasworks was upgraded to the same type of Glover-West vertical plant. Net costs of manufacturing declined to 26 cents from 49 cents per 1,000 cubic feet – a 47-percent reduction. It was results such as these that prompted similar investments at the Vancouver (1920 and 1925), St. Thomas (1922), Quebec City (1929), and other gasworks.[17]

In the natural gas industry, the production challenge was not so much cost as it was the nature of the resource itself. Tilbury and Turner Valley gas had a repugnant odour caused by sulphur content. To make it more palatable, this gas had to be passed through purification plants, such as those constructed by the Union (1924), Royalite (1925), and Southern (1928) gas companies. Other gas, such as that found in the Haldimand-Norfolk and Viking fields, was entirely odourless. In these cases, the gas required odourization, so that leaks in homes and gas mains could be detected more easily. This was done at plants like the one constructed by Northwestern (1932). Though adding to production costs, both these efforts helped to ensure better and safer service.[18]

Along with lowering costs and improving services, gas companies reached out to new customers. The industrial, commercial, and residential areas of service offered a wide scope for further business development. In his address to the CGA in 1925, AGA president Alex Forward pointed out that each year two of the largest car-manufacturing concerns in his country purchased "as much gas as a town of 10,000 people might have used years ago." Moreover, he noted, "one baking company alone" used as much gas as "the average 500 homes." It was much the same in Canada. In Vancouver, the British Columbia Electric Power and Gas Company's gas engineer, John Keillor, reported that the volume of two recent commercial heating orders amounted to that of his company's entire business ten years before. At Northwestern Utilities, the volume of commercial and residential sales increased by 147 percent while revenues from these fields increased by 237 percent between 1924 and 1928 alone. Industrial volume and sales rose by 237 percent and 170 percent, respectively, in the same period as well. And still there were thousands of consumers who might yet be converted to the use of gas.[19]

ALONG WITH LOWERING COSTS AND IMPROVING SERVICES, GAS COMPANIES REACHED OUT TO NEW CUSTOMERS

How were these sales to be achieved? Part of the answer was in the founding of industrial sales departments, such as the one created at the Consumers' Gas Company. Its purpose was to secure sales among manufacturers, hotels, bakeries, and other similar businesses. Beginning as a single-desk operation in the early part of the twentieth century, it gradually came to encompass an appliance showroom, a sales force, and a service crew by the early 1920s. The King Edward Hotel, the University of Toronto, and the *Toronto Star* were numbered among its clients.[20]

19: As gas companies reached out to new customers in the commercial and residential markets, the idea of the gas showroom took hold – a controlled venue for displaying cutting-edge appliances and technology. The Victoria Gas Co.'s showroom, circa 1926, is pictured. *Intercolonial Gas Journal*

20: The gas-rich Turner Valley site was at the forward edge of gas technology from the 1920s onward, employing the first sour-gas scrubbing plant, completed in 1935. *Turner Valley archive*

HOME-SERVICE DEPARTMENTS COVERED RESIDENTIAL GAS SALES, ESTABLISHED BETTER CUSTOMER RELATIONSHIPS, POPULARIZED AND SOLD GAS APPLIANCES, AND INCREASED SALES

21: Part cooking school, part marketing initiative, the value-added home-service events attracted large crowds. This photo shows a well-attended inaugural demonstration in Saint John, New Brunswick. *Intercolonial Gas Journal*

WITH THE PROMISE OF PRIZE DRAWS AND THE OPPORTUNITY TO LEARN NEW RECIPES AND COOKING TECHNIQUES, CONSUMERS WERE ENTICED TO ATTEND COOKING DEMONSTRATIONS WHERE THEY WERE TOLD OF THE VIRTUES OF 'COOKING WITH GAS'

21

"Home-service" departments covered residential gas sales. As outlined by Marcella P. Richardson of Ottawa Gas, their chief aims were "to establish better customer relationship[s]; to popularize and sell gas appliances; [and] to increase gas sales." Gas "cooking schools" were one of the ways they did so. These were not unlike modern-day infomercials. With the promise of prize draws and the opportunity to learn new recipes and cooking techniques, consumers were enticed to attend cooking demonstrations where they were told of the virtues of "cooking with gas." Such schools were initiated by Consumers' (1925), Ottawa Gas (1928), Canadian Western (1929), and other companies in the late 1920s and early 1930s. In addition to generating new business, they provided one of the small but growing areas of employment for professional women who were trained as nutritionists.[21]

Other sales methods were even more direct. In Vancouver, Keillor explained how his company had created a card index for keeping "a complete record of the appliance use" in every house, commercial building, and apartment block. Then a special sales campaign for appliances such as water heaters was put on, in which those who were found to be lacking the appliances in question were "bombarded with advertising followed up by a personal call from the salesman." Gone, then, were the days of "order-taking."[22]

Safety first:
The AGA-CGA 'Blue Star' seal of approval

Much like the gas industry in general, the CGA continued to expand. Toward the end of the 1920s, the organization had a single honorary member, 25 company members, 34 employee members, and 76 associate members, for a total membership of 136. Its annual revenues had grown to $5,400 and assets had increased to $2,000, up from $910 and $1,500 in 1914.[23]

Throughout this period, the CGA continued to lobby for tariff adjustments in order to lessen the cost of imported coal and gas equipment, as well as to organize conferences and publish the *Intercolonial Gas Journal*. The association also did important work related to the establishment of the AGA's appliance testing laboratory. Constructed in Cleveland, Ohio, in 1927, at a cost of $150,000, it was to be the largest gas laboratory of its kind in the world. With input from the CGA and other gas organizations, this lab was founded to help ensure minimum standards for the safety, efficiency, and durability of gas appliances. Once these appliances were approved by the lab, the manufacturers could apply for a "seal of approval" issued by the AGA or the CGA, depending on the market in which they planned to sell their product. These seals were displayed on the appliance and employed in corporate advertising. By 1928, according to its director, R.M. Conner, the lab had already tested approximately 30 percent of all gas boilers, 30 percent of gas-fired furnaces, 50 percent of water heaters, 60 percent of space heaters, and 75 percent of all gas ranges. One year later, the CGA had issued more than 100 certificates of approval to Domestic Gas Appliances of Montreal, Clare Brothers of Preston, Ontario, Hotstream Heater Co. of Cleveland, and others.[24]

Within a few years, the work of the AGA laboratory was further broadened to gas applications, gas processing, and pipe corrosion. After years of advocacy by the CGA and other business associations on the need for more domestic research, the federal government also moved in 1928 to establish its own fuel research laboratory under the auspices of the Department of Mines. By collaborating with the AGA and the federal government, the CGA was thereby able to put their members in touch with much larger research resources than would otherwise have been the case. These were sound initiatives in a solid decade of progress. Both were much needed by the industry and the country – because things were about to get a lot tougher.

22: The CGA seal was affixed to gas appliances that met association standards for efficiency, durability, and safety. *Intercolonial Gas Journal*

Tell Your Gas Consumer What This Seal Stands For

YOUR customers should be made acquainted with the fact that over 280 gas appliance manufacturers have met the requirements of the American Gas Association governing the basic standards of safety. Nearly all of these companies have gone well above the requirements in order that consumer protection might be the first consideration under almost all circumstances.

Quite a few of these companies, including nearly all of the Canadian appliance manufacturers, have applied for and are using the Laboratory Approval Seal of the Canadian Gas Association.

The Canadian Gas Association Laboratory Approval Seal on a Gas Range, Water Heater, Space Heater, Central House Heating Unit, Clothes Dryer, Portable Incinerator, Hot Plate, Laundry Stove and Rubber Tubing is a guarantee of compliance with basic national requirements for safety, and **any such gas appliance, or equipment, that cannot meet these requirements** is not fit to be used and all practical means should be employed to prevent its sale and use. Gas companies can aid in creating a wide knowledge of the value of the Seal by advising customers as to its real significance, and asking them to look for it when buying appliances.

For Further Information apply to the Director of Laboratory Approval Division, G. W. ALLEN,

The Canadian Gas Association, 21 Astley Avenue
TORONTO 5, ONTARIO

22

4

Consolidation

Previous: The exuberance of the postwar boom was soon supplanted by unemployment, poverty, and discontent. Members of the Single Men's Unemployed Association are pictured demonstrating on Bathurst Street in Toronto circa 1930. None was spared the hardship, and the gas industry confronted the crisis – adapt or face extinction – with surprising resolve. *Library and Archives Canada*

1: Taken from the Belleville Gas Works, these charts show the dramatic effects a new vertical plant and new infrastructure had on output. The chart measures pressures at the plant according to time of day. The lower chart was made before the improvements; the top after. The 'shape' of demand remains largely the same, peaking at noon; but the pressure needed to meet that demand is greatly reduced and more evenly distributed. *Intercolonial Gas Journal*

The mid- to late 1920s were almost too good to Canadians. The promise of John A. Macdonald's National Policy seemed poised for fulfillment once more. Solid crop yields, rising stock prices, and the furious pace of construction and investment encouraged this view, as did the sense that this renewed era of prosperity just might continue indefinitely. It was this optimism that had convinced Canadians their young nation could sustain no less than two transcontinental railways, thousands of small-scale independent wheat farms, and a multitude of marginally profitable mining operations, pulp-and-paper mills, and manufacturing enterprises. It could not.

Looking back, one can see that warning signs abounded. With a domestic market of only 10 million people, Canada depended on robust external trade as one of its main engines of economic growth. But after several years of acceleration, the international economy started to sputter during the later 1920s. The problems began with higher tariffs and other trade barriers among the world's major economies, each seeking to maintain its relative economic position. This soon triggered an unstoppable chain reaction: the demand for exports evaporated; prices for wheat, newsprint, steel, and other products collapsed; and the slide toward the Great Depression began as these developments reverberated throughout the domestic and international economy.

ALTHOUGH THE ECONOMY DID BEGIN TO RECOVER AFTER THE MIDDLE OF THE DECADE, THE REST OF THE 1930S WOULD BE LEAN YEARS FOR GAS COMPANIES

The Depression dealt a heavy blow to farmers, workers, and investors throughout Canada. In Saskatchewan, the province most dependent on the export of wheat, the average per-capita income fell by a staggering 72 percent between 1928 and 1933. There and in other parts of the country, unemployment soared to as much as one-third of the labour force, as manufacturers cut production, consumers reduced spending, and railway construction ground to a halt. The share prices of blue-chip companies such as Bell Telephone, Ford Canada, and International Nickel plummeted by as much as 90 percent.[1]

The gas industry could not escape these larger economic forces. Plants that had reduced their production quotas needed far less gas and coke, and consumers seeking to save money often did so by turning down the heat and using less hot water. Others could no longer afford to pay their utility bills at all. Between 1929 and 1933, sales of natural gas in Canada declined to $8.7 million per year from $10.2 million, while sales of manufactured gas and its by-products dropped to $29.9 million annually from $39.9 million. And although the economy did begin to recover after the middle of the decade, the rest of the 1930s were lean years for gas companies.[2]

The industry would have to overcome that challenge even as it faced extensive structural changes. Over the course of the 1930s, the future of small- to medium-sized gas manufacturers, which formed the bulk of the CGA's voting membership, was again thrown into doubt by hard times and competition. At the same time, natural gas companies, which were restricted to associate membership status in the association, demonstrated a resilience that belied early skepticism, despite the inroads made by the electrical utilities, the problem of public relations, and the regulatory impulses of the state. In this context, the CGA had to consider some changes of its own.

Contraction and survival:
Manufactured gas in the Great Depression

Like other industrial sectors, gas manufacturing was hit hard during the economic collapse of the early 1930s; unlike many others, however, it did not fully share in the recovery of the later 1930s. In 1929, there were forty-three coke and gas plants in Canada, thirty-five of which provided illumination and fuel services and eight of which produced coke for various industrial purposes. At mid-decade, the number of plants fell to forty-one, including twenty-five illumination-and-fuel plants, eight coke plants, and eight plants producing Pintsch oil gas for lighting railway cars. Then, in 1938, the number of plants climbed back to forty-four. Twenty-seven of these provided illumination and fuel, and the rest produced coke and other related by-products. It was not until this time that the industry finally recovered to the same volume and value of sales it had attained at the end of the previous decade.[3]

Why did this sector fare so poorly? Economic depression and price competition were certainly two of the big factors. But, just as often the problem was that smaller gas plants had become increasingly vulnerable to being absorbed by natural gas and electrical utilities, as the latter extended their distribution systems. This is precisely what happened in communities such as London, Ontario, where Union Gas acquired the City Gas Company in the early 1930s to convert this city to natural gas,

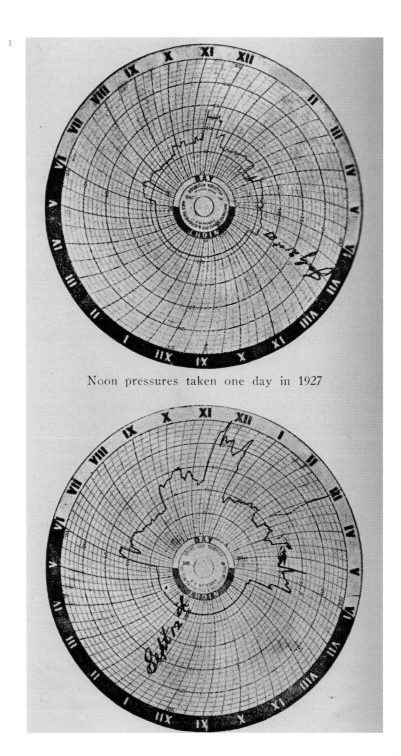

Noon pressures taken one day in 1927

Pintsch gas and the railway

In the early days of railway travel, little thought was given to the question of passenger-car lighting. In the early to mid-nineteenth century, most trips were short and took place by day. But as railways expanded and competition among them increased, so too did the interest in providing for the comfort of passengers over longer journeys, as these journeys began lasting into the late hours of the evening or overnight.

Although coal gas was in widespread use by this time, it could not be practically adapted to the needs of railway companies. Coal gas had to be compressed in order to be transported, causing it to lose most of its illuminative power. This forced railways to rely on candles and kerosene lamps.

In the early 1860s, a Berlin-based tinsmith and lamp maker by the name of Julius Pintsch (1815-1884) found a better solution. He developed a process for refining liquid naphtha into a high-grade oil "distillate," which could then be compressed and stored while losing only a fraction of its illuminative power. Since Pintsch not only invented this process but also supplied the lighting fixtures that usually went with it, "Pintsch gas" and "railway lighting" came to be practically synonymous.

By the late 1880s and early 1890s, the Pintsch system was being employed by railway companies throughout Europe and North America. It reached its height in the first decade of the twentieth century, after which electricity began to make inroads for lighting "luxury cars." As the costs of electrical systems continued to decline over the next three decades, Pintsch lighting was gradually squeezed out of the lower end of the market as well. Declared obsolete by its American manufacturer in 1937, the Pintsch system remained in use on some lines through the lingering Depression of the late 1930s.

Sources: "The Manufacture of Pintsch Gas," *Scientific American*, 9 July 1898; and John H. White Jr., "A Perfect Light is a Luxury: Pintsch Gas Car Lighting," *Technology and Culture* 18, 1 (1977)

and in Cobourg, where the Ontario Hydro-Electric Power Commission purchased the town's gas plant in 1930 only to shut it down several years later, offering to supply its former gas customers with electrical ranges at subsidized prices.[4]

All was not bleak. Small- to medium-sized gas plants could survive – and even thrive – given the right circumstances. Typically, this called for some combination of geographic isolation from natural gas competition, a reliable base of industrial and domestic consumers, and an ownership committed to the progressive modernization of plant facilities. Such was the case in Guelph. Gas manufacturing there dated back to 1868, when a small group of local residents established a works for the purposes of supplying illumination. Service was initiated in 1871, but the plant was not an immediate success. At $5.00 per 1,000 cubic feet, the company's high prices likely played no small role in this, despite the "generous" allowance offered to new customers in order to compensate for air in the pipes when their services were first turned on. Nevertheless, the plant managed to trundle along, thanks to an early base of customers that included City Hall, the Royal Hotel, and the *Guelph Mercury*, as well as various banks, churches, and other downtown businesses.

In 1903, the municipality acquired the gasworks as part of its wider forays into public ownership for the purposes of promoting civic improvements and boosting the local economy. The takeover by the city enabled the plant's management to invest in a more modern, $65,000 horizontal plant in 1906. The timing was fortuitous, ensuring the plant would survive the arrival of hydroelectric power from Niagara Falls in 1908, which was also supplied through the Guelph Heat and Power Commission. By the early 1930s, however, this plant had also become outdated. But rather than entirely switching over to electricity, the city raised the necessary capital for the construction of a bigger and better gas plant. This decision spoke to the continued competitiveness of manufactured gas, as well as the general desirability of providing consumers with the choice of different types of utility services for meeting different types of need.

Completed in 1932, the new plant was a Glover-West vertical system, purchased from the British West Gas Improvement Company at a cost of $120,000. This was the same design as was in use in Vancouver, Ottawa, Halifax, and other communities across Canada, and for more than half of the manufactured gas sold in Great Britain. With its automated stoking process, the efficiency of the system was such that gas rates in Guelph were reduced from $1.00 per 1,000 cubic feet in January 1932 to $0.75 per 1,000 cubic feet in December 1933 – making them the lowest rates for manufactured gas in North America.[5]

Smaller plants continued to conduct business in the urban centres of Sherbrooke, Peterborough, Kingston, Kitchener, and elsewhere. During the late 1920s and early 1930s, there was also some discussion of combining manufactured and natural gas to produce a "high-grade artificial gas" to meet periods of peak demand and stretch out the supplies of natural gas. Union Gas constructed this type of plant at Windsor. Opened in November 1929, it employed the Dayton process of manufacturing gas from oil, which had been developed by the General Oil and Gas Corporation of New York. Over the next decade, this plant added close to 6 million cubic feet per day to the company's generating capacity. The era of the small gas plant was not quite over.[6]

Even so, the future of gas manufacturing was clearly centred on larger urban communities. By the early 1930s, the gas companies in Toronto and Montreal accounted for more than 70 percent of the manufactured gas produced in Canada, with much of the rest being supplied by the other major plants in Vancouver, Hamilton, Winnipeg, and Ottawa. These were the companies that tended to have the managerial, legal, and financial resources to outwit, outmuscle, or, if all else failed, buy out their competitors.

The Montreal Light Heat and Power Company (MLHP) was a prime example. It was formed in 1901 as a result of the merger of the Royal Electric Company, the Chambly Manufacturing Company, and the Montreal Gas Company. The driving forces behind this venture included Senator J.L. Forget, a Montreal-based businessman and politician; James Ross, a railway developer

2: The efficiency gains made possible by the construction of a new plant in Guelph were enough to make area rates North America's lowest in 1933. *Intercolonial Gas Journal*

3: In order to meet peak demand, smaller companies often stored gas in spherical reservoirs. The Hortonsphere erected in the early 1930s marked the reintroduction of gas to Trois-Rivières, Quebec, after an absence of twenty-four years. *Intercolonial Gas Journal*

and the president of the Montreal Street Railway Company; and Herbert S. Holt, an engineer and the president of the Montreal Gas Company. Initially cool to the idea, the young and ambitious Holt was eventually brought on side with the offer of the presidency of the consolidated enterprise. Like his fellow investors, he also came to perceive the business benefits of reducing inter-utility competition and operating all of the once-separate companies under a single management.

As Armstrong and Nelles explain, however, MLHP still had to address the considerable threat posed by the potential development of the numerous hydroelectric sites located nearby. If brought on line by a rival concern, any one of these sites could easily cut into the new company's profits. Indeed, some of these profits were already being drained away by the Lachine Rapids Hydraulic and Land Company (1895), which had forced Royal Electric to cut its power rates by 33 percent. When MLHP was created, Lachine's principal shareholders were thus extended an invitation to join. They declined, hoping they could hold out for even more.

To counter Lachine and other rivals, Forget had become involved in promoting an alternative hydroelectric supplier in the form of the Shawinigan Water and Power Company (SWPC). Its purpose was to exploit the potential of the Saint-Maurice Valley, whose main customer was to be MLHP. But when it became apparent that SWPC's power would cost more than Forget had expected, MLHP opted to develop its own site at another location in 1902. At the behest of its other investors, SWPC struck a deal to deliver its power to Lachine instead. Forget, who had previously lent $50,000 to SWPC, resigned from its board of directors in protest. Then, just as Forget and Holt were planning their next move, their new hydroelectric development at Sainte-Thérèse was suddenly swept away in a massive flood and their Chambly plant destroyed. Most painfully of all, MLHP was forced to buy power from SWPC in order to maintain service.

The MLHP syndicate was undeterred. Though perhaps not quite as fortunate with the weather, they were far richer than their rivals. In early 1903, they again approached the owners of Lachine with the question of what it would cost to get them to go away. The answer was $4.5 million. A hefty price, but one that MLHP's owners decided was worth it. By April of that year, they had acquired Lachine, together with its generating station, distribution system, and power contract with SWPC. Several months later, the directors told shareholders the company could expect to "realize considerable benefit" from the "economical management" of "all the gas and electricity used in the city of Montreal." MLHP had achieved its desired monopoly in Canada's largest metropolis. Around the same time, parallel responses to competition unfolded in other big cities, such as Winnipeg and Vancouver, where the Winnipeg Electric Street Railway Company and the British Columbia Electric Street Railway Company (BCER) each came to control the local gas, electric, and street railway utilities in its respective community.[7]

A somewhat different situation prevailed in Hamilton. Manufactured-gas services were initiated by the Hamilton Gas Light Company (HGLC) in 1850, and for many years it remained the exclusive gas supplier. This changed in 1904, when the Ontario Pipeline Company (OPC) began supplying parts of the city with natural gas from the Haldimand field. Following a decade of stiff competition, HGLC was bought by its competitor in 1913. Under normal circumstances, this probably would have meant the end of gas manufacturing in Hamilton. Natural gas was, after all, less expensive than its artificial counterpart; but with the high level of local demand and the tenuous state of the natural gas supplies in Haldimand, OPC figured it could not afford to abandon gas manufacturing completely. Indeed, in 1923 it decided to construct an even larger plant, which would operate as the Hamilton By-Product Coke Ovens Company (HBCO).

Union Gas was eyeing the potentially lucrative Hamilton market as well. In 1930, it purchased United Fuel Investments, a holding company that controlled HBCO and the former properties of OPC. In addition to gaining the Hamilton franchise, this transaction brought the chance to strike back against an old rival. At the time of Union's takeover of United Fuel, the Dominion Natural Gas Company was supplying the Township of Barton, portions of which had been annexed by the City of Hamilton. Because Union had bought the rights to distribute gas in "Hamilton proper," Union officials felt this was their territory, even though the original franchise did not cover this area.

Revenge is always a bad basis for business decisions, and this was to be no exception. In the next two years, Union convinced city council to "clarify" the company's position as Hamilton's exclusive provider of natural gas. Union then initiated a court action to challenge Dominion's right to supply the annexed portions of Barton. Despite losing its case in the lower courts, Union persisted in pursuing the litigation to the highest levels. In 1934, Canada's top court finally ruled that Dominion's franchise for the annexed portions of Barton remained valid, although Union could also compete in these areas. As longtime gas industry journalist Victor Lauriston points out, this decision resolved nothing. Having created duplicate distribution systems and services, "neither company could look forward to making much money in the contested area of the city. By the same token, neither company was in a position to just pack up and leave." It would not be until 1938 that a "businesslike solution" was finally reached, with both Union and Dominion agreeing to supply a specified portion of natural gas to the annexed areas. Thereafter, the rest of Hamilton would be supplied with manufactured gas by United Fuel under a joint Dominion-Union ownership agreement.[8]

Yet perhaps the most remarkable story of survival in gas manufacturing in the 1930s was that of the Consumers' Gas Company. Located in Toronto, Consumers' was situated next door to the Ontario Hydro-Electric Power Commission's (HEPC) monumental hydroelectric power development at Niagara Falls, which offered some of the lowest electrical rates on the continent. Since it was a private company, its ownership structure offered absolutely no protection from this public competitor.

More, as a public utility, the HEPC was completely exempt from federal, provincial, and municipal taxation, an advantage that Consumers' understandably characterized as "brutal and unfair." Under these circumstances, it is little wonder that Consumers' lost money every year from 1930 to 1938. It is surprising that it was able to carry on at all. Understanding how it did so calls for a closer look at gas markets and strategies.[9]

Finding the right niche:
Canadian gas markets in the 1930s

If the relative price of energy products was the first half of what determined their marketability, the relative quality of the services they provided was the other. Energy markets were not solely dependent on the local conditions of supply and demand. They were also dependent on the comparative reliability and efficiency of the machinery and appliances that could harness the different forms of energy. Customer service counted as well. Each of these factors had a powerful influence on the size and scope of the various energy sectors, hence on the revenues of energy companies.

From the perspective of gas companies, industrial consumers remained among the most sought-after segments of the energy market. Gas had a wide array of industrial applications, and industrial consumers placed the largest and most stable orders. In addition to providing higher sales, such orders improved the efficiency of the entire gasworks by evening out the peaks and valleys in the daily and seasonal demand for gas. The more demand remained constant, the less time expensive equipment sat idle.[10]

Commercial customers were equally desirable for much the same reasons. Hotels, restaurants, dry cleaners, and other commercial establishments all used gas, employing it for everything from refrigeration and air conditioning to hair-drying and heating. Commercial cooking was among the largest of these applications. Studies conducted by the CGA and AGA in the early 1930s suggested why this was the case: baking with gas was three times less expensive than with electricity. And, despite being almost twice as expensive as coal and oil, gas offered a cleaner, more convenient, and better-quality baking method than other "crude fuels."[11]

4: Despite the Depression, Canadians weren't spared the icy rub of winter, so residential heating became an increasingly vital but complex part of the equation. This home, on Laurier Avenue in Ottawa, was a prototype of the all-gas house used for testing and marketing by the Ottawa Gas Company. *Blue Flame of Service*

4

The mainstay of the gas industry, though, was in domestic sales. In the case of Northwestern Utilities, 63.7 percent of its corporate revenues for 1936 were generated from domestic accounts, with 25.5 percent from commercial and 10.8 percent from industrial accounts. Similar patterns were repeated at other companies. Success in the competition for supplying Canadian homes was therefore vital, forming the front lines in the battle between gas, oil, and electricity, as each sought to seize a larger share of the domestic markets for cooking, water-heating, and other services.[12]

As it had in commercial establishments, gas proved to be an efficient and popular method of cooking in the home. There were fifteen gas ranges sold for every electrical range sold in 1933. In fact, the number of gas ranges sold in 1935 alone exceeded the entire number of electric ranges installed in the previous twenty-five years, a large percentage of which were located in areas where there had been no gas service available. Domestic cooking was the gas industry's market to either defend or lose. This, in large part, explains why the home-service departments in many gas companies continued to expand even throughout the cost-conscious days of the Great Depression. By the late 1930s, these departments employed more than 1,000 women across North America, and the scope of their activities had been gradually widened from the presentation of cooking demonstrations on company premises to making house calls, answering telephone inquires, and providing guest speakers for social clubs and radio shows.[13]

In the water-heating business, gas and electricity were more closely competitive. Given the relative consistency of demand for heated water throughout the year, this business represented an important component of the domestic market. So much so that in 1933 the HEPC took the "extreme measure" of offering "free installation" of small water heaters. For the charge of a few cents per month in rental fees, customers would receive equipment that normally cost between $40.00 and $50.00. In return, they would agree to sign either a "flat rate" or "flat rate with measured service" contract, with HEPC providing power at very low rates in order to build up their business. Gas companies were concerned, but remained confident they could compete. Several responded by stretching the payments for their water heaters over a period of three to five years, so as to bring their monthly equipment prices closer to that offered by the HEPC. The 1933 report of the CGA's Competitive Fuels committee further pointed out that customers with electric heaters could not receive the same quality of service as those with gas heaters without installing an additional power "booster." Nevertheless, this was to be a hotly contested market for a long time to come. That same year, there were 27,717 electric- and 11,707 gas-operated water heaters sold in Canada.[14]

The next big market was climate control. Air conditioning was a very small portion of this business; only the wealthiest customers could afford this luxury in the 1930s. Everyone in Canada, however, needed some form of heating. Most had to make do with wood or coal, but a growing number were beginning to switch to oil or gas. The introduction of "special heating rates" by local utilities accelerated this process. In Vancouver, for instance, BCER reported that, following the introduction of such a rate, its gas heating business increased from 5 percent to 23 percent of its total volume of revenues during the period from 1924 to 1933. Thus, there were certainly many areas in which gas could compete with other energy services. The challenge was in figuring out how to expand the industry and the economy at large in a very tough economic climate.[15]

Combatting the Depression in the public and private sectors

The downturn in Canada's economic fortunes coincided with a change in government. Under Mackenzie King, the Liberals had guided Canada through much of the prosperity of the later 1920s. Aside from efforts to redress regional grievances through adjustments in tariffs, freight rates, and federal subsidies, they introduced few other social or economic policies – a stingy Old Age Pensions Act (1927) being their only major initiative. Then, in April 1930, King meandered into an "uncharacteristic slip" in a parliamentary debate on unemployment. The Conservatives were demanding immediate federal assistance to alleviate the financial pressure on provincial governments, which bore the primary responsibility for financing expenditures on social welfare under the British North America Act. King responded by alleging Conservatives sought nothing more than to subsidize their Tory provincial counterparts in Ontario, New Brunswick, and Nova Scotia, and that he "would not give them a five-cent piece."

This collection of graphs yields interesting results about Vancouver during the 1920s and early 1930s. Growth proceeds more or less unabated through to 1931, with the exception of sales of pipe and appliances, which begin to slide as a result of the crashing economy. Most telling, however, is the graph indicating the number of customers, which increased steadily in spite of the unparalleled economic crisis. The graph on the lower right details a decline in wasted gas, evidence of technical and infrastructural progress.

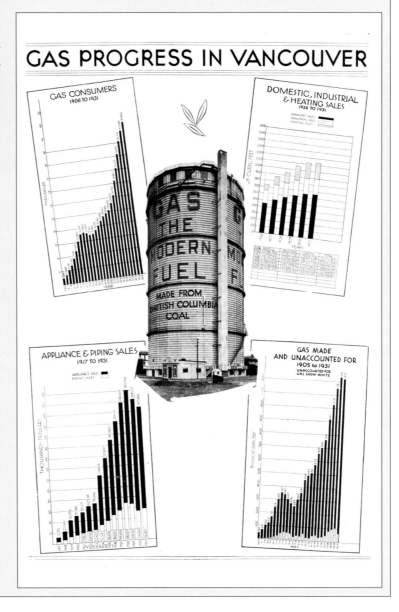

Source: Intercolonial Gas Journal

5: Beating the Liberals in 1933 on a traditional conservative platform, R.B. Bennett did an about-face two years later, proposing liberal reforms to invigorate a stagnant economy and better the lives of Canadians. His 'New Deal' was unkindly met and he was defeated at the polls in 1935. *Library and Archives Canada*

In the ensuing federal election campaign that summer, the Liberals were haunted by King's gaffe and their denials of the severity of the unemployment situation. By contrast, the Conservatives appeared competent, concerned, and decisive. Their leader, R.B. Bennett, was a highly dynamic and successful corporate lawyer with years of political experience, and he vowed to use the tariff to "blast" Canada's way into the markets of other countries. Canadians decided to give him the chance.[16]

Upon assuming office, Bennett raised Canadian tariffs and provided more than $20 million for unemployment relief. Neither measure made a notable impact. Soon the federal government would be spending more than $60 million per year on relief projects alone. As the costs of social assistance increased, so too did the indebtedness of all three levels of government. At the depths of the Depression, in 1933, public charges against the national debt reached an alarming 51 percent of national revenues, while those of the provincial governments reached 33 percent of their combined income. Many municipalities were simply broke. Following the orthodox thinking of the day, these governments reacted by firing civil servants, slashing wages, and tightening expenditures in other areas. It only added to the general misery. In 1935, with another election imminent, Bennett sought a dramatic change of course. After years of emphasis on fiscal conservatism and harsh "law and order" policies, he abruptly proposed to enact a "New Deal" for Canadians, including legislation to establish unemployment insurance, minimum wages, and maximum hours of work. Few were impressed by this "death-bed conversion" to social reform. On 14 October 1935, 45 percent of voters opted for King's Liberals, with 30 percent voting for the Conservatives and 25 percent voting for "third parties" pledging more radical solutions to Canada's economic and social problems.

5

The Liberals moved cautiously once more. This, they explained, was not the time for "untested and unsound" legislative experiments. Bennett's New Deal was referred to the top court, where most of it was ruled to be outside federal jurisdiction, prompting another series of government investigations. Meanwhile, the national unemployment rate slid slowly back to 10 percent – an improvement, but a long way away from the 3-percent rate of the late 1920s.[17]

Within the gas industry, many companies were obliged to undertake serious cost-cutting measures as their revenues declined. These often included reductions in payroll. At BCER, the wages of all employees were reduced by 5 percent or more in the early 1930s (pro-rated on total monthly income). At Consumers', the wages of all employees were rolled back to 1932 levels in 1937. But those who worked for gas utilities were more fortunate than many of their fellow Canadians. They had a job.[18]

6: Despite big hopes for the
St. Lawrence Valley, the 1930s
brought little success for the region
when development picked up in
the latter part of the decade. The
comparisons to Alberta's gas-rich
Turner Valley, pictured, were little
more than wishful thinking.
Intercolonial Gas Journal

Ongoing efforts to expand the gas industry created employment for workers in steel mills, machine shops, and other related businesses as well. Throughout the 1930s, several gas companies embarked on major investments aimed at improving distribution networks and extending services. Union, for example, generated hundreds of jobs with its program to convert London to natural gas, a project that entailed laying a new pipeline, conducting a comprehensive market survey, and refitting countless customer hook-ups. Consumers' upgraded its entire pipeline system from 1928 to 1936 as well, and Northwestern Utilities constructed its odourization plant at Viking in 1932.[19]

The drive to combat the Depression further prompted gas companies to pay closer attention to sales and public relations. CGA and AGA conventions were replete with presentations on improving sales, advertising, customer contact, and collection practices. The common thread was the same: sales development had to be aggressively promoted on all fronts and customers had to be treated with respect.[20]

More growth, more conservation:
The natural gas sector in the 1930s

As the recovery of the later 1930s took hold, the natural gas sector showed signs of revival. By the end of the decade, the value of natural gas sold in Canada climbed to $12.5 million per year: $7.2 million in Ontario, $4.9 million in Alberta, and $200,000 in New Brunswick. The sector employed $79 million in capital and 1,966 workers, compared to manufactured gas's $93 million in capital and 3,930 workers.[21]

AT TURNER VALLEY, ANYWHERE FROM 100 TO 200 MILLION CUBIC FEET OF GAS CONTINUED TO GO 'UP IN SMOKE' EACH AND EVERY DAY

Across Canada, the search for more natural gas went on. During the late 1930s, small amounts of gas were discovered and sold in Saskatchewan and Manitoba. But with a value of less than $31,000, these wells caused little excitement. Instead, some of the biggest hopes for prospectors resided in the St. Lawrence Valley. Farmers there had long reported seeing gas bubbling up from the bottoms of their wells. The resemblance to William Herron's huge discovery at Turner Valley in 1914 was uncanny. With the massive Montreal market situated less than 200 kilometres away, the commercial possibilities of such a well were tantalizing. These possibilities explain why, as one contemporary observer put it, the area was already "studded with the ruins of deceased gas companies which have punched holes all over the map." This did not stop three companies from sinking six more wells in the south shore region. None found any gas.[22]

7: In order to revitalize its flagging Bow Island gas fields, Canadian Western began re-circulating excess gas from Turner Valley into depleted reservoirs to great effect. The pressure more than doubled. *Intercolonial Gas Journal*

8: As president of the AGA in 1933, former CGA president Arthur Hewitt (1910-12) unveiled the 'Fountain of Flame' in front of Gas Industry Hall at the Century of Progress Exposition in Chicago. *Intercolonial Gas Journal*

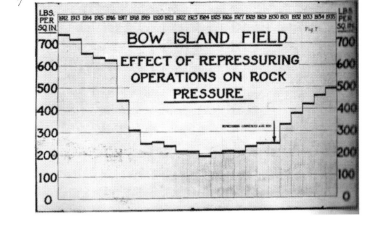

7

At Turner Valley, anywhere from 100 to 200 million cubic feet of gas continued to go "up in smoke" each and every day. Proposals to pipe this gas elsewhere on a large scale kept circulating with little effect, as did plans for liquefying the gas for distribution to "prairie farms and small rural centers." The main practical effort to address this situation was undertaken by Canadian Western, which wanted to use this gas for "recharging" its depleted Bow Island field. By the late 1920s, the practice of storing natural gas by pumping it back into exhausted fields was being successfully employed by a number of companies in the United States. After "exhaustive investigations" of both these reservoirs and the geological structures at Bow Island, the plan went ahead. Between 1930 and 1935, more than 7.9 billion cubic feet of gas was saved as a result.

This still left a great deal of waste. For years the provincial government could do very little about it. This was because the federal government still maintained control over the lands and natural resources of the three Prairie provinces, ostensibly for the purposes of promoting immigration and settlement. But since most of the arable land in the region was settled in the period leading up to the First World War, this policy was to become a growing point of contention, until the federal government relented in 1930.

The Alberta government then moved to take action. Under the United Farmers of Alberta (UFA), it passed the Turner Valley Gas Conservation Act (1932). This act created a three-member conservation board to investigate and control the use of gas in Turner Valley. The board's mandate was to limit the waste of natural gas in the interests of conservation, while allowing for the most profitable naphtha wells to remain in production so as not to unduly harm the economic development of the region. Naphtha producers challenged the act, arguing that the conservation board was an unwarranted infringement of their existing "vested rights." In *Spooner Oils Ltd. et al. v. the Turner Valley Conservation Board* (1933), the Supreme Court of Canada agreed, ruling that if such an infringement was to be retroactive, the intention of the legislature had to be more specific. Together with the political backlash generated by the many interests connected with the development of Turner Valley oil fields, this decision caused the UFA to rethink its position. There would be no appeals. Five years would pass before suitably redrafted legislation put forth by the Social Credit government, elected in 1935, finally managed to regulate the field.[23]

Table 4.1: Ottawa Gas Company cost comparison for house heating requiring 174,000,000 BTUs per season at Canadian prices in 1930

Fuel	Price per Unit	Quantity	Cost
Bituminous Coal	$9.00/ton	13.2 tons	$118.80
Anthracite Coal	$16.00/ton	11.9 tons	$190.48
Fuel Oil	$0.113/gallon	1726 gals.	$195.04
Distillate Oil	$0.13/gallon	1960 gals.	$254.80
Gas	$0.80/per 1000 cu. ft.	391,000 cu. ft.	$312.80

Costs	Bit. coal	An. coal	Fuel oil	Dist. oil	Gas
Installation	$200.00	$200.00	$1000.00	$650.00	$615.00
Interest (6%)	12.00	12.00	60.00	39.00	36.90
Depreciation	10.00	10.00	125.00	81.25	24.60
Repair + Service	10.00	10.00	25.00	25.00	5.00
Fireman	70.00	70.00	–	–	–
Elec. Current	–	–	2.00	0.50	–
Interest on Stored Fuel	2.00	4.00	3.00	3.30	–
Fuel	118.80	190.40	195.04	254.80	312.80
Total	$222.80	$296.40	$410.04	$403.85	$379.30

Source: Roy Soderlind, "House Heating with Manufactured Gas," Intercolonial Gas Journal, October 1930, 365

Constitutional change: The first CGA revolution

The profile of the CGA was on the rise. It held its first conventions outside of Central Canada, meeting in Halifax in 1930 and in Vancouver in 1936. Its directory of approved gas appliances and accessories was being received by more than 700 "gas utilities, plumbers, dealers, heating contractors, and members" by the end of the decade. And, in 1934, its former president (1910-1912), Arthur Hewitt, of the Consumers' Gas Company, had been elected to the AGA – an unusual accomplishment for a Briton who had spent the great bulk of his career working for a Canadian gas company.[24]

Toward the middle of the 1930s, however, the membership structure of the CGA had begun to show some weaknesses. Under the system created in 1919, manufactured-gas companies were the only "Class A voting members" of the association. This posed two problems. First, it affected the financial state of the association, because the dues of these members were based on the amount of gas sold each year. All other members paid a flat fee. As the revenues of manufactured-gas companies had declined, so too had the revenues of the association. Secondly, since natural gas providers made up approximately one-third of the total gas industry, their absence from more formal participation in the association was becoming increasingly awkward for an organization that purported to represent "the gas industry."

THE PROFILE OF THE CGA WAS ON THE RISE — IT HELD ITS FIRST CONVENTIONS OUTSIDE OF CENTRAL CANADA, MEETING IN HALIFAX IN 1930 AND IN VANCOUVER IN 1936

A membership revolution: The CGA's constitutional changes of 1936

Under the strain of financial pressures and a changing membership base, the CGA embarked on a major revision of its constitutional structure in the mid-1930s.

According to the revised constitution, the new classes of membership would be:
  A: Gas Utilities (or public service companies and commissions)
  B: Officials of Class-A members
  C: Associate Members, including individuals or companies directly or indirectly interested in the advancement of the gas industry, except consulting engineers
  D: Individual Members, including individuals who have an interest in the advancement of the gas industry, except consulting engineers
  E: Honorary Members, including persons of accomplishment and service in the gas industry

Membership fees would be assessed at:
  Class A
    • Manufactured-gas companies: Minimum charge of $25.00, plus 12.5 cents per million cubic feet of gas sold in the preceding year
    • Natural gas companies: Minimum charge of $25.00, plus 7.5 cents per million cubic feet of gas sold in the preceding year
  Class B
    • $5.00
  Class C
    • $20.00
  Class D
    • $3.00

And, the affairs of the association would be managed by:
  an executive committee composed of the president, the two vice-presidents, a secretary-treasurer, and ten active members of the association, of whom at least eight must be delegates of Class-A members, or members of Class B, while the remaining two may be delegates of classes A or C, or members of Class B.

  Either delegates of Class A or members of Class B could hold the offices of president or vice-president, and the terms for these offices were limited to two years for the president and three years for vice-presidents.

These changes, therefore, represented a substantial restructuring of the association. For the first time, natural gas companies, gas appliance manufacturers, and individuals "interested in the advancement of gas" could fully participate in the affairs of the CGA. The great majority of the association embraced the new constitution, recognizing the strength that was derived from greater co-operation and collective action.

1948

5

Resurgence

Previous: Canadian Western's service fleet, circa 1940.
ATCO archive

1: Canada declared war on Germany on 10 September 1939. Prepared for the Canada Gazette, the proclamation was syndicated by the Canadian Press and published simultaneously in dozens of Canadian newspapers. *Globe and Mail*

1

Proclamation of War

OTTAWA, Sept. 10 (CP).—Following is the text of the proclamation published to-day in an extra edition of the Canada Gazette declaring a state of war exists be-tween Canada and Germany:

TWEEDSMUIR,
(L.S.),
CANADA:

George the Sixth, by the Grace of God, of Great Britain, Ireland and the British Dominions beyond the Seas, King, Defender of the Faith, Emperor of India.

To all to whom these presents shall come or whom the same may in anywise concern,

Greeting:

A PROCLAMATION.
ERNEST LAPOINTE, ATTORNEY-GEN-ERAL, CANADA.

Whereas by and with the advice of our Privy Council for Canada we have signified our approval of the issue of a proclamation in the Canada Gazette declaring that a state of war with the German Reich exists and has existed in our Dominion of Canada as and from the 10th day of September, 1939:

Now therefore we do hereby declare and proclaim that a state of war with the German Reich exists and has existed in our Dominion of Canada as from the 10th day of September, 1939.

Of all which our loving subjects and all others whom these presents may con-cern are hereby required to take notice and govern themselves accordingly.

In testimony whereof we have caused these our letters to be made patent and the Great Seal of Canada to be hereunto affixed. Witness: Our right trusty and well-beloved John, Baron Tweedsmuir of Elsfield, a mem-ber of our most honorable Privy Council, Knight Grand Cross of our most distinguished Order of Saint Michael and Saint George, Knight Grand Cross of our Royal Victorian Order, member of our Order of the Compan-ions of Honor, Governor-General and Com-mander-in-Chief of our Dominion of Canada.

At our Government House, in our City of Ottawa, this 10th day of September, in the year of Our Lord One Thousand Nine Hundred and Thirty-Nine and in the third year of Our Reign.

By Command,

W. L. MACKENZIE KING,
PRIME MINISTER OF CANADA.

With the outbreak of the Second World War, Canada faced another major international conflict. Once again, the gas industry was a vital contributor to Canada's war effort and postwar recon-struction. Gas companies played a part in almost all branches of wartime production, paid millions of dollars in taxes, and employed thousands of workers. In addition, more than one-third of Canadians made use of gas services in their homes. More than ever, the gas industry was part of a new Canada – urban, industrial, and boasting a rising standard of living. This position furnished ever greater opportunities for the industry, along with a growing set of responsibilities for the CGA.[1]

Again to war:
Canada in the Second World War

Like those of the first, the causes of the Second World War were exceedingly complex. The harsh terms imposed upon Germany by the Treaty of Versailles in 1919; the widespread political, economic, and social instability of the 1920s and 1930s; the failures of the forces of moderation in Italy, Germany, and Japan; and the slowness of the western democracies to confront the expansionist designs of these totalitarian dictatorships are commonly cited factors. The hostilities began with Germany's invasion of Poland on 1 September 1939. Britain and France, which had guaranteed Poland's security after witnessing Germany's absorption of Czechoslovakia in March of that year, declared war on Germany two days later.

Britain's declaration of war did not necessarily mean that Canada was at war as well. Having suffered thousands of casualties and deepened internal divisions during the First World War, Canada had moved much more boldly in asserting its independence from Great Britain in the realm of international affairs. At the Paris Peace Conference of 1919, Canada had signed the Treaty of Versailles in its own right and became a member of the newly established League of Nations. In the 1920s, Canadians began fashioning a distinctive foreign policy and dispatching diplomats to other countries. The 1931 Statute of Westminster, which proclaimed the "legislative equality" of Great Britain and its dominions, is frequently seen as the formal recognition of Canadian independence.

But if the British Empire was dead by the 1930s, support for Great Britain and liberal democracy was not. There were still many Canadians who felt a strong attachment to the "mother country," and who justly feared the growing spread of fascism. Most others were at least willing to go along – up to a point – with what was being touted as the national duty. Canada pledged its support for Great Britain and France in a virtually unanimous war vote in the House of Commons on 9 September 1939.

With respect to compulsory military service, which had caused such great friction during the First World War, King promised that Canada would restrict itself to a "limited commitment" to the war effort. This was to include material support by the government, in the form of money, supplies, air-force training, voluntary enlistment in the military by Canadian citizens, and a contingent of soldiers for overseas service in late 1939. During the early months of the war, this policy of limited commitment enabled Canada to assist Great Britain abroad while maintaining national unity at home.

It was not, however, a policy that could be pursued for long. Between September 1939 and June 1940, Germany seized Poland, Denmark, Belgium, Holland, Norway, the Netherlands, and France, leaving Canada as Britain's major ally. The full extent of the wartime emergency was now evident.

Canadians responded with sacrifice and resolve. In April 1940, the federal government passed the Munitions and Supply Act in order to better mobilize the nation's resources for the war effort. The act endowed the minister of the newly created Department of Munitions and Supply, C.D. Howe, with broad discretionary powers to "mobilize, control, restrict, or regulate essential supplies, and to provide for their transformation into necessary war equipment." Three days after the fall of France in June of that year, this legislation was followed by the National Resources Mobilization Act (NRMA), which provided even greater powers to requisition private property and compel enlistment in the military for "home defence" or any other war-related work. By late 1944, there were more than 750,000 men and women serving in the military. Many other Canadians did important work on the home front by putting in long hours at work, purchasing victory bonds, conserving essential goods, and volunteering for everything from fundraising to collecting scrap metal and other materials.[2]

The demands of the war effort brought rapid improvement in Canada's economic fortunes. Increased government spending

Many women are doing men's jobs on gas works "for the duration." Here is one "walking the plank" with a load of bricks.

and recruitment of workers into the military and war industries helped get the business cycle moving at a much better pace. From 1939 to 1943, federal expenditures on "defence and mutual aid" increased to $4.2 billion from $125 million. This brought the entire federal budget of 1943 to a total of approximately $5.2 billion, a little more than all ten of the federal budgets from 1930 to 1939 combined. In this same year, government statistics

IMMEDIATELY PRIOR TO THE WAR, THERE HAD BEEN FEWER THAN 6,000 WOMEN WORKING IN CANADIAN FACTORIES — TOWARD THE HEIGHT OF WARTIME PRODUCTION, THIS NUMBER GREW TO MORE THAN 250,000

revealed there were roughly 1.2 million workers directly or indirectly employed in war-related work, and that the net value of national production had skyrocketed by 167 percent since 1939.[3] Tighter labour markets were one immediate result. Unemployment, which had plagued so many Canadians during the 1930s, shrunk to less than 2 percent. To compensate for the

3: Letters of support such as this one,
   sent to Canadian Utilities workers
   serving overseas in 1943, were often
   published in company literature.
   The Courier

3

MESSAGE OF THE CALGARY OFFICE STAFF TO THE BOYS OVERSEAS

215 - 6th Avenue West,
Calgary, Alberta,
October 25th, 1943

Dear Fellows:

Another year has rolled around, a year filled with great achievements, but greater things at home and abroad must occur ere we are all together again.

Generally, everyone and everything here, is much the same as when you left. Some changes have taken place. A few articles are scarce, but all in all it could be a great deal worse before we can complain:

This has been quite a busy year so far and, due to the scarcity of men, some of those who were on pension have been brought back to help out.

Some new faces are to been seen amongst the feminine section of the staff and although they are quite demure at present, will no doubt become just as saucy as those who have been with us longer.

Most of the boys who are attached to the various Cadet Corps spent their holidays with their respective Corps, the Sea Cadets at Chestermere Lake and the Air Cadets at various R.C.A.F. Stations in the Province.

The Bowling League is again under way and is progressing very well. Here is proven the kindheartedness of our girls. All young and most attractive, they forego the pleasure of young men's company to gather with the old crocks in order that the various teams can obtain a decent score. When the season is over, as you may have noticed by the last issue of the Courier, they again come forth and entertain these same old fellows by riding hobby horses around the Auditorium.

The Junior Foootball season is over locally. Jack Grogan's outfit, the North Hill Blizzards, started out like a bunch of snowdrops. Towards the end of the season, however, they lived up to the name of Blizzards and came through with a sparkling finish to cop the Championship. They have since been to Edmonton and blew the Edmonton outfit off the field by the score of 23 to 2.

Several of the boys in the Navy have been home on leave during the summer and they are all very well and enjoying life at sea.

We had a letter from Newt Gillespie in Sicily. He is now probably helping make life miserable for the Jerries in Italy.

You may have heard that Lt. Col. Jim Jefferson was awarded the D.S.O. Which proves a quiet fellow can do as well, or better, than a lot of noisy hell-raising hombres.

Several of the girls have abandoned single blessedness, and to all intents and purposes, are very happy in their new venture.

The Fifth Victory Loan Campaign is well under way and though the sum to be raised is quite large, all seem confident that it will go over the top.

The Hockey Season will soon be starting but just what will be the set-up here is uncertain. The teams will be from the various services and at present there is quite a bit of shifting going on as players are moved around to various commands. All of which leaves a great many people rather cold towards it all, especially when a far greater game is being played by the fellows Overseas. A hockey league which may be especially favoured is the Pee Wee. The kids may not be very skilful but their hearts are surely in the game and they try so hard in order that they may be the possessor of a brand new sweater. Speaking of sweaters, have you read or heard this little poem?

"A beauty by name Henrietta,
Just loved to wear a tight sweater,
Three reasons she had:
To keep warm wasn't bad —
But her other two reasons were better."

One difference between this war and the last is that a dame dare not challenge a guy for not being in uniform, she might have to explain why she is not in uniform herself.

The whole Company was saddened by the recent passing of Tom Cavanaugh. Tom had been quite ill for some time. He will be greatly missed by all who came in contact with him.

Hoping you are enjoying good health and are high in spirits. Our spirits are very low, at the Vendors, but otherwise we are all O.K.

Sincerely,

THE GANG.

growing shortage of labour, increasing numbers of women were recruited into a variety of "non-traditional" occupations. Immediately prior to the war, there had been fewer than 6,000 women working in Canadian factories. Toward the height of wartime production in 1943, though, this number grew to more than 250,000, as women came to be employed as welders, riveters, machine operators, and general labourers. Many women assisted the war effort by taking up part-time labour in, as one wartime bulletin put it, "essential services such as hospitals, restaurants, hotels, laundries and dry cleaning establishments."[4]

THE COMBINED PAYROLL OF CANADIAN COMPANIES JUMPED BY MORE THAN 160 PERCENT DURING THE WAR

Wages rose as labour became increasingly scarce. "For many Canadians mired in dead-end jobs," as historian Peter McInnis notes, "the war years proved to be an opportunity of a lifetime .... Across the country people simply walked away from low-paying work and into something better." In British Columbia, an estimated 25 percent of workers in war-related industries were earning $36 per week by 1942, which was 50 percent higher than the pre-war average and "$2 per week more than the traditional trend-setting occupations of the Pacific Coast logging and railway-running trades." In other parts of the country, "domestic workers who earned $20 per week doing their thankless tasks could now expect to make that same amount per week in war-related occupations." All told, the combined payroll of Canadian companies jumped by more than 160 percent between 1939 and 1943, prompting the Bank of Canada to remark that "the average standard of living has risen materially and was probably higher than it had ever been."[5]

Responding to 'the present emergency':
The gas industry in the Second World War

The resurgence of the Canadian economy was immediately reflected in the nation's gas industry. As R.M. Conner, director of the AGA testing laboratories, recounted at the CGA's annual convention of 1941, gas services were being employed in "each step in the production of tanks, guns, airplanes, and ammunition," where it was used "in operations such as melting, annealing, case hardening, carburizing, and any one of the thousands of different heat-treating and chemical processes." In fact, according to one 1942 survey of the fifteen largest gas utilities in Canada, gas was used in as many as 530 war-related industries and 58 "camps, barracks, and military hospitals."[6]

The domestic side of the industry was equally vibrant. Many Canadians who had been crowded into multiple-family dwellings during the 1930s began to move into their own homes and apartments, while others migrated from outlying rural areas to take up better jobs in urban areas

4: To say the gas industry was instrumental in the Allied victory, as this 1943 political cartoon humorously predicts, is no stretch. From gas-fired munitions plants to gas-fired cook tops at home and abroad, the industry was a key contributor to the international war effort. *Gas World*

4

The Menu of Mars!

The extent to which gas appliance makers are adapting their activities to war munitions production was indicated by Mr. Arthur Forshaw in a recent address

(Cartoon by Wallace Coop)

Cumulative Waste!

Mr. Coal : Well, ain't there a war on ?
Mr. Gas : All the more reason for stopping this business !

[*Cartoon by Wallace Coop*]

served by gas utilities. As housing vacancies went down, the number of gas meters in service went up. By 1941, there were more than 700,000 gas meters in service.[7]

Sales of gas appliances were strong as well, at least until the introduction of a 25-percent excise tax in 1941, a policy that was specifically intended to discourage the manufacture of domestic appliances in order to conserve precious metals and other materials for the war effort. In a further push to conserve the outflow of foreign currency, except for war purposes, a prohibition against importing "all gas apparatuses designed for cooking and the heating of buildings, as well as gas refrigerators and air-conditioning units" was imposed as well. The Wartime Prices and Trade Board (WPTB) further ordered that "no old gas ranges could be scrapped without government approval." At a meeting of the CGA's executive committee in 1943, it was noted that one unintended consequence of this ban was that some of the untrained WPTB inspectors had allowed old gas ranges back into service that were, in the opinion of gas companies, unsafe. The matter was promptly taken up with the WPTB by CGA president Frank D. Howell, who would later report that "the matter was now better understood and most likely the situation complained of would not now occur again."[8]

Despite these hurdles, domestic gas services continued to play an important role in Canada throughout the war years. Gas was still the most efficient method of domestic cooking, a fact bolstered by continual improvements to the gas stove. In the early 1940s, tests conducted by the AGA indicated "the average efficiency of gas range top burners" had increased by about 50 percent in the past fifteen years, while that of gas-fired water heaters had increased by 25 percent. What's more, home-service departments provided suggestions on how "to conserve food, and preserve fruits and vegetables, as well as [how] to select and use home grown products." And appliance manufacturers, such as Servel Incorporated of Evansville, Indiana, furnished Canadian companies with "literature, posters, and advertising material on nutrition, at less than cost," so as to keep workers healthy and productive.[9]

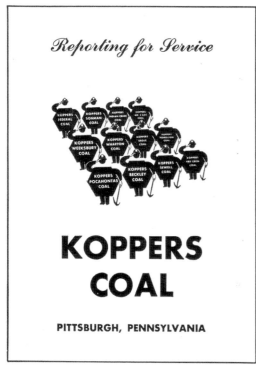

Reporting for Service

KOPPERS COAL

PITTSBURGH, PENNSYLVANIA

6

7

5: Although the war abroad often demanded co-operation among competitors at home, the battle for market share raged on. The waste of resources was tantamount to treason, so British gas companies used the spirit of the Conservation Act to take a dig at their 'inefficient' competitor, coal. *Gas World*

6: One of a series of highly recognizable Koppers Coal wartime advertisements, circa 1941. *Intercolonial Gas Journal*

7: According to tests carried out by the American Gas Association in its world-class laboratories, pictured circa 1942, the gas stove in the 1940s was 50-percent more efficient than it had been fifteen years earlier. *Gas World*

IN THE EARLY 1940S, TESTS CONDUCTED BY THE AGA INDICATED GAS WAS STILL THE MOST EFFICIENT METHOD OF DOMESTIC COOKING

Gas companies struggled to respond to the rapidly changing conditions of wartime Canada. At Northwestern Utilities, for instance, the business of war-related establishments accounted for 20 percent of the total gas load by 1943. These customers included the Royal Canadian Navy Volunteer Reserve, the No. 2 Air Observers School of the Royal Canadian Air Force, the No. 4 Initial Training School of the RCAF, as well as US contractors working on the Alcan and Canol projects, US Army Engineers in the Dominion Motors Building, and Northwest Airlines. The creation of "war work" also brought a large influx of new workers, so that between 1939 and 1943 Edmonton's population jumped to 105,536 from 90,419 (or 16.7 percent in just three years). Together, these developments brought the "maximum peak hourly demand for gas" to 326,000 cubic feet, up from 15,000. Already stretched to the breaking point, the company then learned that Edmonton's war housing project was in the process of planning the construction of an additional 250 housing units and "contemplating the installation of gas in each [of these homes] for domestic purposes."

Meeting these demands required ingenuity. When it became apparent the company would not receive the pipe it had ordered in time for the 1942 construction season, owing to greater wartime priorities elsewhere, the company nonetheless managed to make good with co-operation, a bit of luck, and a massive reshuffling of its existing stock of pipe. Company president Julian Garret described the process of addressing the rising volume requirements in Edmonton:

> We took up 9,400 feet of our duplicate 12-inch line, and replaced it with used 16-inch pipe which we were fortunate in obtaining from another company. We then enlarged our line between Viking and Kinsella by taking up 4.1 miles of 6-inch pipe and replacing it with the same amount of 12-inch pipe, using the pipe recovered from the main line and the pipe left over from the previous year. The 6-inch pipe which had been so removed, was used to lay field lines to our new wells.[10]

The work proceeded on schedule, and winter demand was met.

Comparable strains existed elsewhere. In 1941, the Montreal Coke Company (MCC), which supplied gas for the MLHP and coke for domestic heating, was unable to meet the full demand for coke as a result of greater-than-expected government orders for cooking and heating in the local military camps. In the following year, MCC was forced to devote its entire production to wartime purposes. Around the same time, the brisk expansion of heavy industries in Hamilton obliged United Fuel Investments Ltd. to restrict the sale of coke from the Hamilton By-Product Coke Ovens plant to "the immediate territory" of the city, and construct a $3.1-million coke-oven gas plant with financial assistance from the federal government. Even these measures could not prevent shortages.[11]

Coal was scarce throughout Canada. Toward the end of the war, the situation became so serious the government decreed that "no one should employ in a non-mining job anyone who had mined coal since 1935." Because of the lower heating value of Canadian coal and the costs of transportation, most gas companies in Central Canada used imported coal. In return for the privilege of doing so, these companies had to pay a 10-percent War Exchange Tax on such imports. Through this measure, the federal government was able to raise an estimated $1 million per year for the duration of the war. Moreover, by creating not only manufactured gas but also coke and other by-products from these imports, gas plants contributed to mitigating the nation's larger fuel and chemical supply problems.[12]

Gas company employees lent substantial assistance to the war effort, too. With skilled labour at a premium, many older gas workers – some with as many as forty years of service – continued to carry on throughout the war, even though, as Consumers' president Edward J. Tucker recognized, their "retirement was long overdue." No doubt this must have been particularly difficult for those charged with the tough labour of coal handling, well drilling, and pipe laying. Nevertheless, many of these and other workers still found time and energy to play a part in civilian defence work, blood service, and further war-related activities.

Out of a workforce of about 6,000 people, more than 800 gas workers participated in overseas service. Jack Price of the Canadian Utilities office in Indian Head, Saskatchewan, was one of them. He served as a pilot officer (in this case, a navigator) in a bombing squadron that took part in "numerous large-scale raids over France, Holland, Denmark, and Italy." It was a very dangerous assignment. For the aircrews of the British Royal Bomber Command, which included Canadian pilots such as Price, the statistical likelihood of surviving the war "physically unscathed" was only 24 percent.

Gas workers kept up with the news of their enlisted cohorts through reports published in company magazines. The details of one of Price's missions were so recorded. In the summer of 1943, his bomber was attacked by an enemy fighter "after blasting [the German city of] Stuttgart." Three of the four engines were put out of commission and the plane lost height rapidly.

CANADIAN WESTERN STAFF OVERSEAS CIGARETTE FUND

By M. C. JAMES

M. C. James

This project has now been in operation for a little over two years and during that period eighty-four thousand and nine hundred cigarettes have been despatched to the gas-house gang who are now serving overseas.

At its inception there were approximately eighty office staff members contributing twenty cents per month in order to despatch twelve cartons of cigarettes overseas each month. Thus, for a time, it was possible to build up a reserve. The peak load has been twenty cartons per month, while the number contributing,

GAS COMPANY STAFF CIGARETTE FUND

due to enlistments and retirements, has decreased from eighty to an average of fifty each month.

COME TO ☿ YOUR UNCLE

and be
SAVED

Use your CREDIT and end up in the "Dog-House."
Our TERMS are SO REASONABLE you'll say WE are PHILANTHROPISTS.

TERMS:
15 days or less............10%
30 days or less............20%
All profits provide bounty for "the boys"
COME EARLY AND AVOID THE RUSH

Your friend and benefactor,
MICHALOVITCH JAMESTEIN

"THE DIME TOBACCO AND LOAN ASSOCIATION"

In order to augment the fund, without raising the subscription rate, it was decided to form a subsidiary and to advertise its function by means of the above card. This venture, we are glad to report, has been a howling (especially on repayment day!) success, and has been honoured by the distinguished patronage and generous co-operation of all great and earnest spendthrifts on the staff.

We believe this fund has proved its worth as despite other Attractions (for single men only, oh yeah!) the boys still pay court to Madame Nicotine if the many cards of thanks we receive are a sign of her continued popularity.

In some cases delay has been caused by incorrect addresses — and notification of transfer to other units would eliminate this trouble.

Geo. Benoy tells us that at least one of his shipments was lost through enemy action but, on the whole, we have been lucky in this regard.

We were pleased to hear recently from a number of the boys and an Airgraph from Mark Talbot was much appreciated.

"KEEP 'EM SMOKING"

—— (C) ——

8: Many staff members of Canadian Western donated twenty cents per month to a cigarette fund, which sent approximately 85,000 cigarettes to fellow 'gas-house men' serving overseas. *The Courier*

9: Northwestern Utilities could barely keep up with wartime demand as it was, when a contract for 250 gas-equipped war houses in Edmonton brought consumption to new heights. *The Pilot*

10: Jack Price of Indian Head, Saskatchewan, was one of roughly 800 workers from the Canadian gas industry who served overseas.
The Courier

10

As told in the *Courier*: "Over France the crew stood by to bail out. Over the channel they prepared to crash in the water. However, with the skill of the pilot, the flight engineer's ability to balance gasoline tanks, and the jettisoning of equipment, the bomber [managed to] reach England at a height of 7,000 feet." It crash-landed on a Midlands airdrome and Price went on to survive the war. More than 45,000 Canadians were not as fortunate.[13]

Shortages of raw materials and labour were only a couple of the challenges that gas companies faced. During the war, the average costs of coal advanced by 40 percent and the average costs of labour rose by 35 percent. Gas companies were also faced with investing additional monies in protecting their facilities against possible sabotage or "air raids," as well as training staff on how to prepare for such emergencies. On top of these rising costs, more taxes had to be collected and paid. With the costs of the war mounting, the federal government imposed an additional 8-percent tax on gas sales and raised the minimum corporate tax rate to 40 percent. Publicly owned utilities remained largely tax exempt.[14]

In spite of all this, the gas industry made some promising advances during the early to mid-1940s. During the last three-quarters of 1940, Union posted a 35-percent gain in operating profits over the same period of the previous year. Its partially owned subsidiary, United Fuel, nearly tripled its earnings in the middle of that year. On 20 December 1942, Consumers' Gas registered its largest single-day volume of sales in its ninety-four year history, with an output of 22.859 million cubic feet. Over the course of the war, the entire Canadian gas industry served 5.5-percent more customers and sold 18.7-percent more gas.

At the CGA convention of 1946, Tucker assessed the performance and position of his industry. He explained that the gas industry's ability to meet the considerable stresses provoked by the war was largely owing to the "almost universal adoption by member companies of the industry of the forward-looking policy of being prepared at all times to meet greater demands for its service. Its production and distribution facilities were in excellent condition with ample margin of reserve plant." Building on this head start, gas companies continued to keep pace with increasing demand "by operating those plants beyond their rated capacities and by keeping them in continuous operation through the deferring of much maintenance work." Now there was much work to be done to reconstruct and renew both the nation and the gas industry, and to meet the public's expectations for a better standard of living and "a higher standard of service." The planning had already begun.[15]

11: Perhaps nothing so clearly signifies the importance of gas in modern society than this aerial photograph of a German gasworks razed in an RAF strike. *Gas World*

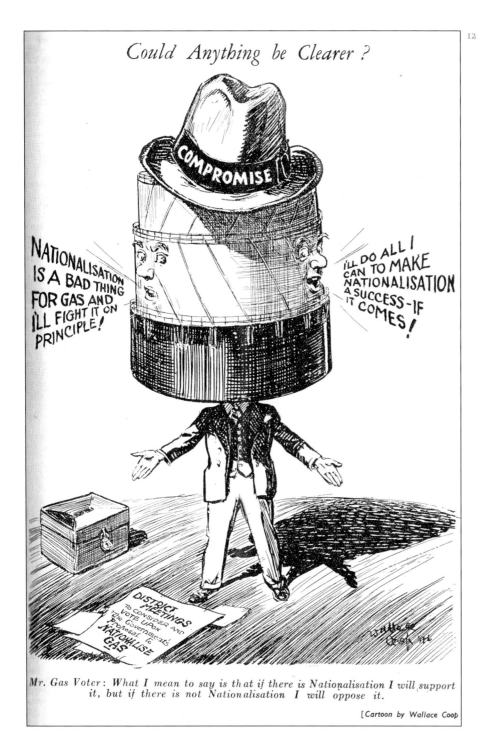

Could Anything be Clearer?

COMPROMISE

NATIONALISATION IS A BAD THING FOR GAS AND I'LL FIGHT IT ON PRINCIPLE!

I'LL DO ALL I CAN TO MAKE NATIONALISATION A SUCCESS—IF IT COMES!

DISTRICT MEETINGS TO CONSIDER AND VOTE UPON the Government's Proposal to NATIONALISE GAS

Mr. Gas Voter: *What I mean to say is that if there is Nationalisation I will support it, but if there is not Nationalisation I will oppose it.*

[Cartoon by Wallace Coop

The transition to peace:
Postwar planning and reconstruction

The belief that the Allies were fighting for a better world sustained many Canadians through the most difficult periods of the war. In a national broadcast on government plans for war and peace given at London, Ontario, on 30 May 1945, Prime Minister Mackenzie King recognized the importance of this motivation in declaring that Canadians had "always professed to regard the war as a struggle for freedom and that, in truth, is what it was .... But the final defeat of the attempt by Germany and Japan to dominate and enslave the world will only be the end of one more battle in the age-long struggle for man's freedom." In the postwar world, he asserted, freedom from the "fear of war" would have to be accompanied by the freedom from the "fear of unemployment."

During the same broadcast, King pinpointed what many regarded as the essential trade-off in securing the latter freedom. "If Canada's war effort has been the magnificent success it has," he noted, it was "because every aspect of Canada's war effort has been carefully planned, organized, and directed. Following the end of the war, the achievement of full employment and social security will require no less careful planning and wise direction." However, he added, "in peacetime, the planning and direction of government must be exercised in such a way as to increase, and not decrease, the opportunities for individual freedom and initiative." It was, therefore, King's aim "to reconcile the maximum amount of individual freedom with the greatest possible measure of mutual aid in the attainment of human welfare."[16]

Canada's participation in the war effort had certainly enlarged the federal state's potential capacity for "careful planning and wise direction." During the wartime emergency, state regulation of prices, wages, labour conditions, and a host of other areas of economic and social activity expanded to unprecedented proportions, as did the administrative apparatus of the federal bureaucracy. At the end of the First World War, there were about

41,800 employees in the national civil service, and in the following two decades this figure edged up to slightly more than 46,100. Between 1939 and 1945, the number more than doubled, reaching 115,900. Among these new recruits were numerous experts from the university, professional, business, and labour communities dedicated to assisting the government in planning for a more stable and prosperous postwar world. By late 1942, several government agencies and committees had already been charged with ensuring that this was the case, and over the next three years postwar reconstruction would become the theme of numerous official investigations and reports.[17]

Leonard Marsh's Report on Social Security for Canada was among the most widely circulated of these reports. Commissioned in December 1942, the Marsh report was intended to follow up on the release of a popular British report with a similar purpose that had been drafted by William Beveridge, a high-level adviser to the British government and the former director of the London School of Economics. As a former student of Beveridge, a founding member of Canada's farm-labour Co-operative Commonwealth Federation (CCF) party, and a former professor of economics at McGill University, Marsh came to the task with the right credentials to help the King government deflect criticism it was not taking an active enough role in preparing for postwar reconstruction.

AT THE END OF THE FIRST WORLD WAR, THERE WERE ABOUT 41,800 EMPLOYEES IN THE NATIONAL CIVIL SERVICE, AND IN THE FOLLOWING TWO DECADES THIS FIGURE EDGED UP TO SLIGHTLY MORE THAN 46,100. BETWEEN 1939 AND 1945, THE NUMBER MORE THAN DOUBLED, REACHING 115,900

For Beveridge and Marsh, the "five giants" on the path to postwar reconstruction included "Want, Disease, Ignorance, Squalor, and Idleness." During the late nineteenth and early twentieth centuries, the effects of these economic and social problems had largely been addressed through a combination of self-reliance, private charity, and limited public assistance. In this same period, though, countless government investigations, university research, and non-governmental organization reports had documented the persistence of social problems and the many inefficiencies of existing methods. What Beveridge and Marsh advocated, therefore, was the adoption of a more co-ordinated and scientific approach, including more government spending on public works, universal health insurance, family-allowance benefits, improved pensions, and other programs. Though their proposals were more ambitious than what the King government was prepared to pursue, they reflected the growing faith in government planning that had emerged out of the wartime experience.[18]

12: Eager to avoid a postwar malaise, economies the world over contemplated various policy options to chart a smooth transition. In Britain, nationalization of the gas industry was one particularly controversial and complex option. *Gas World*

The CGA's manufacturers' section, est. 1947

The safety and efficiency of gas appliances and equipment remained crucial to building the residential, commercial, and industrial gas markets in the 1940s and 1950s. Recognizing the importance of this aspect of the industry, the CGA had been involved with the American Gas Association's appliance-testing facilities since the late 1920s, and made constitutional revisions to allow appliance and equipment manufacturers a greater role in the association during the mid-1930s.

By the mid- to late 1940s, there were dozens of appliance and equipment manufacturers in the CGA. At the annual convention of 1945, one member raised the question of establishing a specific manufacturers' section. Its purpose would be to enable these members to get together at conventions to discuss issues relevant to their businesses. After two years of investigation and planning, the section was officially founded at the CGA's fortieth annual convention. The manufacturers' section later went on to make significant contributions to the creation of the CGA's national appliance-testing facilities and the initiation of the association's popular gas measurement school, along with many other aspects of the gas industry at large.

Sources: CGA archive, Executive Committee, Minutes, 26 Oct 1945; and "Presidential Address of Alex Mackenzie," Canadian Gas Journal, 1 July 1949, 172.

Ambitions and uncertainties:
Postwar planning in the gas industry

The Beveridge and Marsh reports arrived at a time when the discussions on postwar planning and reconstruction were gathering momentum. At the end of the First World War, the transition to peace had been a disaster, beset by wild swings in the business cycle, mass unemployment, and general social unrest. There was a strong determination to avoid a repeat performance. Between 1942 and 1945, scores of studies on postwar reconstruction began appearing in the form of newspaper, magazine, and journal articles, as well as pamphlets, conference proceedings, and books. These studies embraced diverse approaches and organizations, encompassing technical analyses by organized interests, such as the Canadian Life Officer Association's "Public Health Aspects of the Proposed Health Insurance Scheme for Canada" in the *Canadian Journal of Public Health* (April 1943); surveys and recommendations by interested experts, such as

J.A. McNally's study, "The Place of Forest Management in Postwar Rehabilitation Plans," in *La Fôret québecoise* (1943); and popular forums by public and private institutions, such as the Canadian Broadcasting Corporation's *Of Things to Come... [An] Inquiry on the Postwar World* (1943).[19]

Gas associations were involved in these discussions. Postwar planning was seen as critical not only for assisting North American society in negotiating a successful transition to peace, but also for capitalizing on the wartime gains in the gas business. The AGA began its work in this area in early 1943. Representatives from the CGA joined it later that year, and the CGA established its own committee to look at the specific conditions within Canada. The membership of the CGA committee suggests the importance of the matter. It included distinguished members of the association such as John Keillor (BC Electric), one of the originators of the CGA; E.C. Steele (Union Gas), a founding member of the NGPA; and Julian Garret (Northwestern Utilities), the president of the CGA from 1939 to 1940. J.D. Von Maur, the distribution engineer at Consumers' and a future CGA president, chaired the committee.

The joint AGA-CGA committee for postwar reconstruction considered a wide range of issues. What were the potential markets for gas services? What factors would affect the realization of these markets? What was the gas industry's capacity to meet future demands? And what would be the influence of long-term national trends? The conclusion was that gas had a significant role to play in meeting North America's future energy needs, particularly in industrial and commercial heat applications and domestic cooking and heating. To secure this place, the committee resolved that the association had to undertake a more active program of national advertising and do a better job of encouraging the insulation of commercial buildings and domestic residences.[20]

Though broadly supportive of this program, the CGA struggled with the question of what could be done within Canada. The association recognized there was a strong possibility of a slump in gas sales during the immediate postwar period. With the war over, the demand for anti-aircraft shell cases, machine-gun tripods, and other military equipment manufactured with gas would largely disappear. These factories would either have to be reconverted to producing something else or be closed down, and gas companies would have to look elsewhere to recover this lost load. The CGA maintained that advertising was one important aspect of the larger strategy for promoting future expansion, yet it was unable to secure sufficient support for its plan to implement a $20,000 postwar national advertising campaign. Indeed, its proposal for a 4-cent-per-meter "tax" to raise the necessary funds received support from companies representing only 16 percent of the membership's active gas meters – a figure that fell quite short of the association's 70-percent target. The nation's thinly spread population and regional differences, Tucker surmised, presented too many barriers to the success of such a program. For the time being, the association simply encouraged its members to pursue more local initiatives.[21]

Throughout the 1920s and 1930s, gas companies had begun to devise and adopt increasingly sophisticated sales methods. Appliance showrooms, gas exhibitions, and home-service departments were all part of these efforts. In the 1940s, the trend continued. It was at this time that BC Electric, a longtime pioneer in the promotion of gas services, introduced comprehensive customer surveys to ascertain the preferences and tastes of their customers. Some were "arranged by telephone," but most were "straight cold canvass." The "girls" conducting the surveys were "instructed to get into the kitchen where possible ... [to] obtain the information," and leave behind a booklet on how customers might "get the most" out of their gas or electric range. These surveys revealed that those who favoured gas ranges appreciated "speed" above all other factors, while those who favoured electric ranges leaned toward "cleanliness." They also indicated that the gas ranges of no fewer than 200 out of the 1,411 customers surveyed required some sort of adjustment. Key selling points were thus identified, along with areas in which the company could improve the satisfaction of existing customers.

The CGA and the International Gas Union

Numerous international organizations were founded in the late nineteenth and early twentieth centuries. Their functions varied, from providing a forum for establishing common standards, to furnishing a means for resolving international disputes, to disseminating information on administrative and technological developments from around the world. The International Telegraph Union (1865), the Universal Postal Union (1874), the League of Nations (1919), and the International Labour Organization (1919) date from this period. They were created with the hope that greater international co-operation would foster improved economic efficiency, national security, and better living standards for all.

Formed as a non-governmental, non-profit organization in 1931, the International Gas Union (IGU) was part of this trend. Registered in Switzerland and headquartered in Denmark, it was created to provide an international counterpart to the various national gas associations that constituted its membership. Much like these associations, its objectives were to promote the use of gas services and allow for an exchange of information and ideas among members.

As in other areas, however, engendering international co-operation in gas services was a slow and halting process. Although among the early supporters of the IGU, the CGA was not entirely convinced of its value. With the focus of its concerns centred on developments in Canada, the United States, and Great Britain, the CGA's executive committee was dissatisfied with the information furnished by the IGU in the late 1930s, much of which dealt with France, Germany, and other continental European nations. The rising cost of membership, which had increased to 500 from 200 Swiss francs by the mid-1940s, was another consideration. As a result, the CGA's membership lapsed in the early 1940s. But it was reinstituted in 1946 during a resurgence of internationalism after the Second World War.

Sources: CGA archive, Executive Committee, Minutes, 5 June 1939–27 Nov 1950.

GAS COMPANIES HAD BEGUN TO DEVISE AND ADOPT INCREASINGLY SOPHISTICATED SALES METHODS. APPLIANCE SHOWROOMS, GAS EXHIBITIONS, AND HOME-SERVICE DEPARTMENTS WERE ALL PART OF THESE EFFORTS

13, 14: The CGA decided that the best approach to the postwar years ahead was to keep the industry relevant to customers. Advertising at the local and national levels became a dominant initiative, and member companies took up the call by building highly visible, full-service showrooms in the late 1940s and 1950s. *CGA archive*

13

14

The appearance of cartoon-type character as advertising logos was another innovation. In the busy and crowded world of the postwar media, attracting the attention of consumers was becoming harder and harder. For Consumers', the answer was "Speedie Flame," an odd-looking humanoid with stick-figure arms and legs, over-sized shoes, and a giant flame – complete with large eyes, pointed ears, and a big nose – in place of its head. It was difficult not to notice Speedie on billboards and in newspaper advertisements. By way of these and other initiatives, gas companies fought to win their share of the anticipated postwar "sweepstakes" in consumer spending.[22]

As it happened, the dreaded postwar recession never materialized. During the war years, the federal government had introduced a series of measures designed to sustain consumer demand. The Unemployment Insurance Act (1941) provided income security for workers displaced by the closing of wartime factories. The War Service Gratuities Act (1943) provided income for returning soldiers to allow them time to find new work. The Family Allowance Act (1945) provided mothers with a small monthly income for improving the lives of their children. Optimistic investments on the part of private businesses, along with fifteen years of pent-up, deferred spending resulting from the Depression and the Second World War further smoothed the transition to peace. So too did a healthy international environment, in which the United States generously contributed to the reconstruction of Europe through the Marshall Plan, while allowing those countries receiving such aid to spend it in either Canada or the United States. For its part, Canada outright forgave more than $1 billion in British war debts. Everything came together at once, and the postwar economy climbed to unprecedented heights.

This was good news for the gas industry. In Winnipeg, the demand for gas and coke was so great as to keep its works in operation around the clock. There and in other communities, such as Victoria, propane-air plants were introduced to provide a source of "peak-load relief" for the gas plants. As Alan H. Harris, the manager of the Winnipeg Electric Company, explained, "You all know that to get the last drop of juice out of an orange takes a great deal of squeezing. [And it] takes a lot of squeezing under severe weather conditions to get the last 200,000 cubic feet per day out of our coke plant, and this squeezing is expensive, and, in addition, is hard on the equipment." Innovation therefore allowed for the continuing expansion of demand; and as the 1940s drew to a close, the gas industry unremittingly posted new records in sales.[23]

One difficulty that remained was the persistent shortage of natural gas to serve southwestern Ontario. The drilling programs of Dominion and Union Gas had fallen short of supplying wartime needs, forcing both companies to adopt propane plants of their own in the early 1940s. At the 1944 convention of the Natural Gas and Petroleum Association of Canada, geologist John Reeves called the problem "the most pressing need of our industry." That year, Union negotiated an

The CGA and the British Institute of Gas Engineers

The gas industry had deep roots in Great Britain. In 1863, the founding of the British Association of Gas Managers marked the creation of one of the first gas associations of "gas men" anywhere in the world. Many of its early members were engineers, such as its first president, Thomas Hawkesley, who had supervised the construction of more than 150 waterworks and several gas works in Great Britain and other countries. As engineers began promoting the professionalization of their discipline in the late nineteenth century, the association's name was changed to the British Institute of Gas Engineers (BIGE) to better reflect the "professional" basis of its membership.

In the late nineteenth and early twentieth centuries, Canadian businesses often imported ideas, capital, and personnel from Great Britain and the United States. This frequently led to the development of close ties between Canadian, British, and American institutions, as in the case of the warm relationship between the BIGE and the CGA. Members of the BIGE often took part in CGA conventions, and vice versa. On one such occasion in 1933, a visiting delegation of the BIGE presented a silver tray inscribed with the name of then-CGA president Hugh McNair. The tray was presented to each incoming head of the association – a tradition that continues to this day.

Sources: A.E. Haffner, "Centenary Presidential Address," Gas World, 18 May 1963, 586-98 and 608; and CGA archive, "Notes on CGA's Relationship with the Institution of Gas Engineers, for S.A. Farwell," 7 May 1993.

I'm
SPEEDIE FLAME
of
THE CONSUMERS' GAS COMPANY

agreement with the Panhandle Eastern Pipeline Company to import a monthly 400-million-cubic-foot "supplementary supply" of gas from Texas. Political approval and deliveries began shortly thereafter.[24]

After the war, the long-unrealized potential for bringing Alberta's gas to Ontario began to look viable. Technologically, it was now feasible. Since the time of the first gas discoveries in Turner Valley, there had been major advances in gas distribution, and several long-distance pipelines crisscrossed the United States. Geologically, it also made sense. Following a massive oil discovery at Leduc, Alberta, in 1947, investigations by the federal government had established that sufficient provable reserves existed to make a transcontinental pipeline economically viable. But politically it was not yet certain how the financing, terms, and conditions of the export of gas from Alberta to other parts of Canada would proceed. The Social Credit government of Alberta was unsure of how best to protect provincial interests, while the Liberal government at the federal level was unsure whether an all-Canadian or a Canada-US route would be in the nation's best interests. Before making any major changes in plans, most gas companies decided to adopt a wait-and-see attitude. But they would not have to wait for much longer.[25]

A national association for a national industry

The CGA emerged from the 1930s and 1940s as a more confident and nationally oriented association. To reflect this, the *Intercolonial Gas Journal*, the CGA's official organ, was renamed the *Canadian Gas Journal* in 1939. Throughout this period, it had provided a valuable liaison between the federal government and the gas industry, and had kept the membership of the association in touch with one another and with the world. After the war, the CGA had even secured a curious but cost-effective agreement allowing timely statistics to be collected by the AGA on behalf of the Canadian industry, so that Canada's companies could plan for the future.

15: After the war, Speedie Flame
became the cartoon spokesman
for Consumers' Gas, appearing in
billboard, newspaper, and
trade-journal advertisements.
Canadian Gas Journal

16: In Britain, Mrs. and Mr. Therm
were the icons of the gas
industry. Unlike Speedie Flame,
however, the cartoons were used
to promote the industry at large.
Gas World

The growing self-assurance and national perspective of the association was further evidenced by its sudden insistence that the NGPA drop the words "of Canada" from its name, so there could be no possible confusion of the primary "national" gas association. This campaign was initiated by CGA president Julian Garret in his address to the CGA's annual convention of 1940. Garret had a point. The CGA was the "oldest gas association in Canada." It had the broadest membership and the largest network of national and international contacts; and with the constitutional changes adopted in 1936, both manufactured- and natural gas companies played a full part in its affairs. The NGPA, by contrast, was mainly an Ontario-based organization of natural gas companies. Its annual conventions were held mainly for the purpose of discussing matters of "local interest." Garret, therefore, suggested the word "Ontario" take the place of "Canada." The matter would be taken up with the NGPA several more times over the 1940s. With gas poised for a massive national expansion, the CGA had served notice it had every intention of maintaining its position as the national gas association.

6

Transformation

TRANS CANADA PIPE LINES LIMITED

GENERAL MAP

Previous: The prospect of a national pipeline transformed Canadian politics into a maelstrom of debate. With US global influence on the rise, an all-Canadian pipeline was an imperative shared by many. Labor Progressive party representative Harry Hunter is pictured at a demonstration, circa 1951, in the Toronto area. *Library and Archives Canada*

1: Spanning from Burstall, Sask., to Montreal, the $370-million pipeline transformed the gas industry in Canada from a fractured, regional affair to a national one. *Canadian Gas Journal*

In an event that has been compared to the driving of the last spike on the Canadian Pacific Railway, the final welds on the TransCanada pipeline were completed near Kapuskasing, Ontario, in the early evening of 10 October 1958. This time, however, there were no dignitaries and no ceremonies. Just a cold, wet group of contractors working to finish their job so as to get themselves out of another of northern Ontario's miserable sleet storms. The achievement was not even reported in the nation's largest newspapers, the *Globe and Mail* and the *Toronto Star*.

During the autumn of 1958, there were good reasons for the parties involved not to draw attention to the pipeline. It had been initiated by a Liberal government but completed under a Conservative regime. Its creation had been controversial, and its construction was an important reason why the Liberals now sat on the opposition side of the House of Commons. It was better for everyone to allow the moment to pass quietly.

But the completion of the pipeline surely was a significant event. Then the longest pipeline in the world, it was a feat of politics,

engineering, and finance. Stretching from the gas fields of Alberta to the gas markets of Central Canada, its construction involved multiple layers of government, crossed hundreds of miles of rough terrain, and cost more than $370 million. When the first 300 million cubic feet of gas began flowing, on 27 October 1958, the pipeline transformed Canada's natural gas industry from a regional to a national concern. It meant the immediate restructuring of existing gas markets and the emergence of new regulatory frameworks, and it would eventually contribute to making natural gas an even more crucial component of the nation's domestic energy supply and international exports.[1]

The CGA would be transformed as well. As a national gas industry emerged and expanded in the postwar era, the CGA's organizational structure proved less and less adequate for carrying out the association's growing list of economic, promotional, and regulatory functions. Having been designed for the prewar era of regional business models and minimal government, the CGA began the 1950s with few regular staff and a limited capacity for conducting policy research and public relations on behalf of the industry at large. Up to this point, it had achieved much with these resources. Yet if its success was to carry over into the postwar era – characterized by rapid economic and social changes, wide-ranging and persistent public debates, and broad and complex government regulations – then the CGA had to reinvent itself. During the 1950s, its members did so by structuring the association's constitution to allow for the expansion of its membership and resources. These adjustments would gradually enable the association to become something more than a conduit for the collection, filtering, and dissemination of industry developments and membership ideas. They would enable it to become an increasingly effective contributor to the gas industry in its own right.

The coming of the pipeline:
Issues, actions, and consequences

The proposal to build a large-diameter pipeline to export Alberta's natural gas to markets in Central Canada raised myriad questions about the optimal management of natural resources and the appropriate role of government. In the late 1940s and early 1950s, it remained undetermined whether or not such a pipeline should or would be undertaken – even if it was known that it could be done. Nor was it obvious what part the various governments might play in the financing and regulation of the project.

Alberta, for one, was initially cool to the idea. Following the discovery of much larger natural gas reserves in the late 1940s, the Social Credit government of E.C. Manning appointed a commission to investigate the desirability of exporting Alberta's gas. Chaired by R.J. Dinning, the Alberta Natural Gas Commission heard from forty-nine expert witnesses, municipal officials, gas company representatives, and interest groups at hearings in Edmonton, Calgary, and Medicine Hat. Its report, issued in March 1949, recommended that there should be no exports of natural gas until at least fifty years' worth of provable reserves had been secured for the citizens of the province.

The Manning government had to proceed with caution, as Albertans had come to expect low-cost natural gas service. In their representations to the Dinning commission, the Edmonton Chamber of Commerce, the Union of Alberta Municipalities, and the Medicine Hat branch of the Canadian Manufacturers' Association (CMA) all urged that "no steps be taken that would endanger the supply of gas to the people of the province as a convenient and low-priced fuel and as an incentive for the establishment of new industries and businesses." Because landlords and homeowners had spent substantially on domestic heating equipment and gas appliances, municipal officials from Edmonton, Calgary, and Lethbridge wanted to guarantee a thirty- or forty-year supply so as to ensure they would recoup this investment. Farmers, too, had come to rely on low prices. In rural communities throughout southern and central Alberta, it was not uncommon for farmers to run gas lines into their barns to keep their farm equipment, cars, and animals warm and comfortable through the province's long and cold winters.

2: The significance of the pipeline may not have been grasped by Parliament and the popular press; but for those in the industry, the feat signalled a new epoch. To put the achievement in context, the Canadian Gas Journal devoted a special issue to the pipeline. *Canadian Gas Journal*

2

PIPELINE REPORT

A SPECIAL JOURNAL FEATURE

Drawing of Pipeliner by Malcolm Bell, Trans-Canada Pipe Lines Ltd.

CANADIAN GAS JOURNAL

3, 4: Editorial cartoons by artist Robert
   Chambers, published in the Halifax
   Chronicle Herald, show the extent
   to which the 1957 election was
   fought on the issue of the pipeline.
   Library and Archives Canada

THE PIPELINE WAS ONE REASON WHY THE LIBERALS NOW SAT ON THE OPPOSITION SIDE
OF THE HOUSE OF COMMONS

5: When the agreement to proceed with the pipeline was signed on 7 June 1956 – TransCanada president and general manager Charles S. Coates, pen poised, second from the left, is pictured among the signing party – few could predict the wide-ranging effects bringing gas from the West to Central Canada would have on the industry and the country at large. *Library and Archives Canada*

5

Nevertheless, the gas reserves of Alberta offered economic opportunities that were too big to ignore. Prominent Canadian geologists, such as G.S. Hume of the Geological Survey of Canada and T.A. Link of the Northwest Natural Gas Company, assured the commission that more than 4.2 billion cubic feet (or thirty years' worth) of proven reserves could be counted on, and that much more was likely to be found. Many other expert witnesses and gas company officials concurred, and emphasized that the existence of export markets would promote a level of exploration generating more than enough reserves for the province's long-term domestic, commercial, and industrial requirements. Moreover, as still other witnesses pointed out, the potentially huge royalties from gas exports could be used for the benefit of the province.[2]

In light of these considerations, the Manning government enacted legislation designed to strike a balance between the preservation of Alberta's resources and the potential for future exports. Under the Oil and Gas Conservation Act (1949), the province set a reserve requirement of thirty years and created the Alberta Oil and Gas Conservation Board to review applications for export licences. During its first two years of operation, the board denied all such applications. By 1952, however, more than 6.8 trillion cubic feet of gas had been discovered. At this point, the board determined that adequate proven reserves existed to allow for exports from the Peace River district. A permit for these purposes was issued to the Westcoast Transmission Company, which planned to export the gas to British Columbia and the Pacific Northwest of the United States. Soon afterward, it was apparent there were also sufficient reserves for exports from the other producing regions of Alberta.[3]

In the early 1950s, there were two serious contenders for this enticing opportunity. The first group, Western Pipelines, brought together Alan Williamson, a former official of the Department of Munitions and Supply; H.R. Milner, a founding director of

Canadian Utilities; and Canadian investment firms Wood, Gundy and Nesbitt, Thompson. Western sought to export gas from Alberta to Winnipeg and the mid-western United States. The second group, Canadian Delhi Oil, included Ross Tolmie, a well-known Ottawa lawyer; Clint Murchison, a wealthy Texas oil promoter; and the New York-based investment firm Lehman Brothers. Delhi aimed to construct a larger "all-Canadian" pipeline to export gas from Alberta to Central Canada.

From the perspective of the federal government, which was responsible for regulating interprovincial and international trade, neither of these proposals was ideal. Western's plan offered fewer national economic benefits; Delhi's American ownership offered greater political problems. There were also questions of political sovereignty and economic viability. On the one hand, the Western Pipelines proposal would mean Central Canada would continue to be served by natural gas imported from US companies. Could a reliable supply be assured on this basis? On the other hand, the Canadian Delhi proposal called for a far more expensive project. Could the Americans raise the capital on their own or would they look to the Canadian government? And if their ambitious plan were to fail, would taxpayers be left footing the bill? Since both pipelines could not be profitably developed, the solution was a merger of the two initiatives.

The main impetus for that merger came from C.D. Howe, the federal minister of trade and commerce. Howe was an American-born engineer who had made a fortune by constructing grain elevators across Western Canada from the mid-1910s to the mid-1930s. When the Depression ended his business, Howe entered politics. Running for the Liberals in the federal election of 1935, he won the riding of Port Arthur and was quickly appointed to a cabinet position. During the 1940s, as the minister of munitions and supply, he co-ordinated the economic dimensions of the Canadian war effort and, later, as the minister of reconstruction, oversaw the postwar transition to peace. Now, Howe relished the opportunity to take on one last great project. At his urging, the two competing proposals were merged in January 1954. The resulting company,

TransCanada Pipelines, was to have 50-percent Canadian ownership and was to construct a pipeline from Alberta to Central Canada via the "all-Canadian" route, with spur lines leading into the mid-western and north-central United States.[4]

With a viable proposal in place, the next challenge was to secure the necessary financing. Irrespective of its nationalistic merits, the all-Canadian route was enormously expensive. This raised a familiar problem in economic development: how to underwrite a large-scale project with limited capital resources. The Canadian experience with railways suggested three basic options: first, tap into local savings to build small portions of the line over a long period of time; second, encourage private investors to assume the risk in return for future profits; third, have government take on the project through public ownership. As in the case of the transcontinental railway, high costs and time constraints ruled out the first option, whereas pragmatic considerations called for a combination of the other two.[5]

Approximately $375 million would be needed to construct the pipeline, or an amount roughly equivalent to the federal government's entire transfer payments to the provinces and municipalities in 1955. This was too much for the company to raise alone. Therefore, in October 1955 it was agreed that the most expensive portion of the pipeline – through northern Ontario – would be built by a Crown corporation, the Northern Ontario Pipeline Company, under the joint ownership of the federal and Ontario governments. With $118 million removed from the equation, TransCanada would obtain the rest of the money through private channels, and purchase the northern-Ontario section at cost plus interest, once it was able to do so.

The plan did not quite work. Potential investors remained anxious about the fact TransCanada had not secured approval for exporting gas to the United States from the US Federal Power Commission (FPC). Without such approval, many doubted the Canadian pipeline would be a good bet. Banks, steel companies, and other suppliers were equally skeptical about extending credit to such a large and risky venture. As a result, the company began to encounter difficulties even in raising the money that it needed to begin construction of the Prairie section of the pipeline.

To improve TransCanada's prospects, two things were done. First of all, in late 1955, three American companies – Tennessee Gas Transmission, Canadian Gulf, and Hudson Bay Oil and Gas – took a 17-percent ownership position. Although this brought the American ownership content to 51 percent, it also ensured that any pipe purchased by the company would have a use, if TransCanada should fail. Secondly, in May 1956, the federal government agreed to advance $80 million to finance 90 percent of the Prairie section of pipeline. It did so, however, only under the condition that the original private sponsors of the company inject another $8 million into the project (bringing their total investment to $16 million). If the project was not completed by the end of the 1956 construction season, the federal government would assume full ownership of the pipeline and the private sponsors would forfeit their entire investment.[6]

To have any chance of meeting this deadline, TransCanada had to be able to exercise its option on the pipe ordered from American steel mills by 7 June 1956. Working backward from that date, this meant that there was just a little more than five weeks to gain parliamentary approval of the new financing arrangement. It was an extremely tight schedule in the best of circumstances, and in this case the opposition in Parliament was in no mood to co-operate. The TransCanada proposal provided an ideal opportunity to embarrass the government. Public-opinion polls suggested that 45 percent of Canadians favoured a pipeline developed by "private Canadian investment," 29 percent favoured "government ownership," and a bare 17 percent favoured the current set-up of a mixed public and private partnership involving Canadian and US investment. This fit well with the stated positions of the two main opposition parties, as the Conservatives had long argued for "Canadian development," while the CCF had long stipulated "public ownership."

To defeat the government's initiative, all the opposition needed to do was delay the business of the House of Commons long enough for the deadline to pass. There was only one way to avoid this outcome: a parliamentary procedure allowing the sponsor of a bill to invoke the "closure" of debate on a particular issue. Closure had been employed before, but only in cases where debate had dragged on for quite some time. Here, however, there was no time for subtlety. The Liberals took the unprecedented step of invoking closure during the first reading of the bill, before any debate had even taken place. They repeated the same tactic in each phase of the legislative process.

The opposition exploded. "You are the High Executioner of Parliament," bellowed the Conservatives' most theatrical and effective parliamentarian, John Diefenbaker. "Dictatorship!" "The Guillotine!" cried others. Countless questions of privilege, points of order, and motions were introduced, often by the formidable CCF MP Stanley Knowles, an unassuming former United

6: Clarence Decatur Howe, whose diverse cabinet experience earned him the soubriquet 'minister of everything,' was the Liberal trade minister who proposed a merger between contenders for the pipeline. *Library and Archives Canada*

7: Building the world's first 2,294-mile pipeline required specialized engineering and first-rate construction on complex terrain. Many Canadian firms rose to the challenge. Advertisement for Winnipeg-based Price Poole Constructors circa 1957. *Canadian Gas Journal*

7

Church minister who had become renowned as "the Dean of parliamentary procedure" since making the jump to politics in 1942. Each delay cut into the slim timetable.

Forging ahead, the Liberals used their majority in the House of Commons to push the TransCanada bill through the legislative process on time. Since the government was determined, the opposition could not prevent the passing of the bill. What the opposition did do, however, was shift the focus of the debate away from the merits of the TransCanada financing arrangement itself and toward the manner in which was implemented. The central issue looked less like whether this was the best possible agreement that could be negotiated under the circumstances, and more like whether the Liberals had, as Diefenbaker characterized it, "made a mockery of Parliament at the behest of a few American millionaires."[7]

TransCanada did get its loan, but it was not able to complete the pipeline that year after all. Labour disputes in US steel mills prevented the delivery of pipe that summer. This was obviously outside of TransCanada's control, so the private owners of the company did not lose their investment. More than ever, this delay made it appear as if the Liberals had run roughshod over Parliament for no good purpose.

In June 1957, the government sought a renewal of its mandate. By this time, the Liberal party had been in office at the federal level for a total of twenty-two consecutive years. Its leader, Louis St. Laurent, was seventy-eight years old. He had been ready to retire but was convinced to stay on for one last fight. Voters seemed to like "Uncle Louis," as the soft-spoken leader was dubbed. So much so that Fisheries Minister James Sinclair once said his party could win with St. Laurent, even if they had to "run him stuffed."

8: At Kapuskasing, Ont., the last weld bridging the two halves of the TransCanada pipeline was executed without fanfare at 6:46 p.m., 10 October 1958. The era of the pipeline had arrived. *Intercolonial Gas Journal/TransCanada Pipelines Ltd.*

THE FINAL WELDS WERE COMPLETED IN OCTOBER 1958 — THERE WERE NO DIGNITARIES AND NO CEREMONIES — JUST A COLD, WET GROUP OF CONTRACTORS WORKING TO GET THEMSELVES OUT OF NORTHERN ONTARIO'S MISERABLE SLEET

They almost did. Hearkening back to his roots as a corporate lawyer and board member, St. Laurent dryly reported the facts and asked voters to judge his "management team" on their record. In recent years, the economy had grown at an average rate of about 5 percent, unemployment remained below 5 percent, and inflation was largely non-existent. Progress on the Trans-Canada Highway, the St. Lawrence Seaway, and the TransCanada pipeline all counted among the government's accomplishments. As one Liberal MP had put it, there simply were no worthwhile national projects that they were "not already developing."

Such smugness invited questions and comparisons. On closer inspection, it was obvious there were large tracts of the country that had not fully shared in the postwar prosperity of the urban-industrial heartland. In the Maritimes, coal miners and shipbuilders had yet to recover from the loss of their wartime business, while steel workers wondered aloud why the government had not insisted that TransCanada buy its pipe from them. In the West, farmers suffered under some of the lowest wheat prices in years with little federal support, as they watched the government pour money into its other "great projects." More generally, rural Canadians were far less likely than their urban counterparts to have access to basic amenities such as electricity, gas services, and indoor plumbing, not to mention luxuries such as automobiles, radios, and telephones.

The message of John Diefenbaker, the new Conservative leader, appealed to these voters. During the 1920s and 1930s, he had established a reputation as one of Canada's top criminal defence lawyers, often successfully defending underdog clients whom few others would touch. Now he pleaded the cause of the "ordinary Canadian." Over and over, he railed against "Liberal arrogance" and offered up the hopes of the "Conservative vision" of "one Canada, with equality of opportunity for every citizen and equality for every province from the Atlantic to the Pacific." As the historical alternative to the Liberals, the Conservatives were also well positioned as the logical choice for those voters who were just looking to make a change.

But few thought that Diefenbaker could actually win. Pundits talked of his party's landing only another thirty or so seats, and *MacLean's* went to press with an editorial discussing four more years of Liberal government. In fact, the Conservatives more than doubled their representation from 51 to 112 seats in the 265-member parliament, giving them a minority government. Nine months later, they would be swept back into office with the largest majority government in Canadian history.[8]

During the interim, the TransCanada pipeline was completed at long last. Fears of American domination proved to be overblown, as Canadians eagerly bought up its stock and soon comprised the majority of the company's ownership. By the end of the 1950s, Alberta's gas was being distributed by Union Gas in southwestern Ontario and Consumers' Gas in Toronto and northeastern Ontario, as well as Montreal and other parts of Quebec, where it was distributed by the Quebec Natural Gas Company. Numerous other communities across Canada were served along the way.[9]

9: In areas where natural gas was unavailable, propane was an option with a rapidly expanding customer base. This photo of a Liquigas service call was taken in Lachine, Que., circa 1959. *Canadian Gas Journal*

The effects on the gas industry were immediate and dramatic. As natural gas became widely available throughout the most populous regions of Canada, manufactured gas was gradually phased out in favour of this lower-cost and cleaner alternative. Where natural gas services were unavailable, "bottled" propane and liquefied petroleum became another option. The statistics are telling. In the early 1950s, Canada's annual gas consumption was 58 billion cubic feet of natural gas and 27 billion cubic feet of manufactured gas; by the early 1960s, it was 320 billion cubic feet of natural gas and 1 billion cubic feet of manufactured gas. A new era of gas services had arrived.[10]

Toward a national regulatory system:
The founding of the National Energy Board

The evolution of a national gas industry was paralleled by the development of a national gas regulatory system. On the same day that gas started flowing through the TransCanada pipeline, the Royal Commission on Energy tabled its first report. The Borden commission, as it was known, had been established by the Diefenbaker government in October 1957 to investigate the policies that would "best serve the national interests" with respect to the trade, transmission, and regulation of natural gas and oil. The first report, issued on 27 October 1958, recommended that gas exports and imports should proceed; however, it also called for the establishment of a national energy board to regulate the trade and transmission of natural gas, and for the provision of advice to the government on national energy policies. These recommendations echoed the earlier suggestions of a Liberal-appointed commission, the Royal Commission on Canada's Economic Prospects (also known as the Gordon commission), which had reported in November 1957.

ON THE SAME DAY THAT GAS STARTED FLOWING THROUGH THE TRANSCANADA PIPELINE, THE ROYAL COMMISSION ON ENERGY TABLED ITS FIRST REPORT

In light of this emerging consensus, the federal government founded the National Energy Board (NEB) in late 1959. Its responsibilities included issuing licences for gas and electricity exports and imports, as well as making recommendations on the prices and quantities of this trade that would best serve the national interest. The board was additionally charged with providing advice on energy policy and was granted the power to regulate the construction and operation of international and interprovincial pipelines, as well as their "traffic, tariffs, and tolls," by virtue of "its own authority." In the early 1970s, these same responsibilities were extended to include the oil industry.[11]

The CGA provided a forum for the vital questions of the day. A two-part series on "rate regulation" in the *Canadian Gas Journal* during the late 1960s was representative. The first part made the case for the market. Its author, Otto Zwanzig of BC Electric, began by assuring his readers that his argument was "not a plea to eliminate all regulation." Rather, it was a call for the "elimination or relaxation of price regulation in areas where a monopoly no longer prevailed." With the increasingly pervasive accessibility of electricity, oil, and other competing energy services, Zwanzig observed that such areas were few and far between. Less price regulation would permit gas companies to free "top executive manpower" from "their considerable preoccupation with administrative, regulatory problems," and make their attention more available for the problems "of the marketplace." In addition, regulation via the competitive market as opposed to the state would encourage the search for efficiencies that were missing in a regulatory environment governed by

National Energy Board

1984 Annual Report

10

10: The creation of the National Energy Board in 1959 – the board's 1984 annual report is pictured – marked the start of a new era in Canadian energy policy. The CGA responded by refining and expanding its policy and regulatory expertise. *Minister of Supply and Services Canada*

a "cost plus return-on-investment" formula. The result would be higher returns for shareholders and fairer prices for consumers.

The second part laid out the case for price regulation. Citing the "opinion of regulatory authorities," it reiterated the classic argument for the regulation of a "natural monopoly." When competition is restricted for reasons of public interest, high capital costs, or the lack of possible substitutes, as the argument goes, it is necessary for the state to regulate prices in the interest of assuring fair prices. It is neither practical nor desirable, for instance, for multiple telephone or hydroelectric companies to run competing delivery systems for their services. According to the regulatory authorities, gas utilities were in an analogous position, because the majority of their individual consumers could not "quickly or easily dispense with" gas services. If left unregulated, therefore, gas companies could potentially exploit their "monopoly position" at the expense of both the individual consumer and the community at large.[12]

More extensive national regulation placed greater emphasis on the importance of policy expertise in the gas industry and the management of public opinion at the national level. As the models furnished by the Canadian Manufacturers' Association, the Canadian Chamber of Commerce, and other national business associations indicated, success in these realms called for the maintenance of an ongoing research program and an acquired field of knowledge, as well as agreement on a broad and coherent set of policy preferences. In its current form, the CGA was ill-prepared for these roles. Its name appears nowhere among those groups that had made representations to either the Gordon or the Borden commissions, major federal investigations into matters of immediate concern to CGA members. Something had to change.[13]

The second CGA revolution

After the end of the Second World War, the association entered another period of flux. Its membership numbers continued to rise, increasing to 336 in 1953 from 231 in 1947, and its executive-committee minutes remained active, and alive with reports and proposals. Yet the CGA tended to lack the resources it needed for translating its ideas into realities, as is evidenced by its inability to conduct economic research and postwar advertising during the mid-1940s. But there were other deficiencies. The growth of the industry was far outstripping that of the association. From what was once a vital centre of information about international trends in the gas industry, the *Canadian Gas Journal* had gradually become more like a newsletter that reported on recent Canadian events. Moreover, by the early 1950s many of the first- and second-generation CGA "founders" had either passed away or were nearing retirement.[14]

During the first decade of the postwar period, there was an accelerated demand for an effective national gas association. The end of the war brought forth the revival of competition, and the growth of national regulations meant a single national voice for the gas industry was more necessary than ever. Only by speaking in such a voice could the industry ensure that policymakers and the public heard and understood its perspective.

At the annual convention in June 1954, CGA president Dennis K. Yorath pointed out:

> Until recently this association has been a cozy little group of friends, drawn together through business connections, who meet once a year on a more or less social basis and exchange a few ideas concerning their industry. Today, the situation has changed completely .... In my opinion our association has got to become a business organization prepared to get out and do a job on behalf of what is about to become one of Canada's major industries.

He understood that "some of us may regret this change ..., but it cannot be helped." Those in the association could remain "friends," but the "coziness" had to end. He urged that "we must be as realistic as possible and start today to convert our organization into a full-time, efficient body that can assist, advise, and act on behalf of the Canadian gas industry."[15]

Over the ensuing months, the members of the CGA undertook to reorganize the association's structure and increase its financial and human resources. The turning point came with the redrafting of the constitution. Discussions of constitutional changes that had begun among the members of the executive committee in the early 1950s, led to the establishment of a constitutional subcommittee in mid-1954. At the annual convention the following year, the proposals of this subcommittee were put before the members.[16]

11: In June 1954, CGA president Dennis K. Yorath recognized the association was at a pivotal time in its history. In order to survive, the CGA had to cease being a social club and start advocating on behalf of the industry in an increasingly complex regulatory environment. *CGA archive*

Table 6.1 The CGA's membership structure in 1957

As part of the CGA's efforts to expand its base, the association made several further adjustments to membership categories during the postwar period, the most significant of which occurred in the mid-1950s. By the time the TransCanada pipeline was completed, the revamped membership structure was to be as follows:

Class	Membership type	Number of members (1957)
A	Gas Utility Members	31
B	Gas Production Members*	5
C	Gas Gathering, Transmission, or Storage	4
D	Liquefied Petroleum Companies	-
E	Manufacturer Members	132
F	Associate Members**	33
G1	Individual Members – Officials of A, B, C, and D	146
G2	Individual Members – Officials of Class E	108
G3	Individual Members – Officials of Class F	10
G4	Individual Members – Other individual members	12
H	Non-Resident Members	16
I	Honorary Members	13
J	Life Members	7
–	**Total membership**	**517**

* Not marketing to the consumer

** Companies, organizations, or self-employed individuals not covered elsewhere

Sources: CGA archive, Executive-committee minutes, 27 Nov 1953, 3; CGA, Golden Anniversary Proceedings (1957), 10; and CGA, Constitution and Bylaws
of the Canadian Gas Association (CGA: Toronto, 1969).

According to the revised constitution, the number of different categories of membership was expanded to ten [Table 6.1]. New categories included those designated for Gas Gathering, Transmission and/or Storage Members (Class C), Liquefied Petroleum Members (Class D), and Non-Resident Members (Class H), each of which was indicative of the expanding scope of those groups with a vital interest in the gas industry and its association. The status of the manufacturing section was elevated to that of an official and ongoing part of the association, rather than that of a typical subcommittee. Much like the establishment of a separate manufacturers' category of membership, this recognized the growing importance and influence of those members who serviced and sustained the gas utilities and their customers.

The most significant constitutional amendments, however, were those relating to the governing structure of the association. The top governing body of the CGA was to be a board of directors rather than an executive committee, consisting of "twenty-two members, composed of the elective officers (ie. the President and two Vice Presidents), [as well as] the immediate Past President, the Managing Director, and seventeen additional active members of the association." The board's composition was not inevitably to be dominated by the utility members, as it had been since the creation of the association back in 1907. In the new arrangement, the offices of the president, vice-president, and managing director were open to any members of classes A through E, as well as Class G. Furthermore, eight of the seventeen additional board members were to be drawn from "delegates or employees" of the utility class (Class A), five from "delegates or employees" of the manufacturers' class (Class E), and up to four from classes B, C, D, F, or G.[17]

The new constitutional structure was approved by the membership at the annual convention of June 1955, and the association was formally incorporated under its provisions one year later. Other changes soon followed. The constitution provided that the president was to be its chief executive officer, but it further provided that the board of directors may appoint a managing director from time to time. Once appointed, the managing director was to assume "the general management and direction" of the "Corporation's business and affairs," subject to "the authority of the Board and the supervision of the President." Two months after the constitution had been approved, William H. Dalton was appointed as the first of these directors.[18]

The development of the CGA's Laboratory Approvals program was one of Dalton's early responsibilities. After the end of the Second World War, manufacturers of gas appliances and equipment had become "anxious" to make certain that only "approved appliances" were being sold. The maintenance of high standards was vital to the long-term interests of the gas industry. If gas appliances proved to be unsafe or unreliable, few would be willing to purchase gas products or services. Many manufacturers were also concerned about the diverging standards of approval that were slowly taking shape across Canada. In the early 1950s, certain gas utilities in Alberta and Ontario were refusing to accept the CGA/AGA standards, and some provinces had either established or begun to consider their own specific codes. Taken to its logical conclusion, this multiplicity of regulation could have been ruinous to the gas industry. Other

manufacturers were dismayed at the "excessive costs" of sending their products for testing at the AGA laboratories in the United States.

At first, the utility representatives on the CGA's executive committee were hesitant to support the establishment of a Canadian laboratory. The existing facilities in the United States served Canadian needs, they thought, and would not support duplicative efforts. The manufacturers, however, were dissatisfied with the system because some "individual companies made individual demands." At the end of what was described as a "full and frank" discussion of the issues at the executive-committee meeting of April 1951, the CGA president, H.G. Smith of Consumers' Gas, acknowledged that, if the members of the association "accept and proclaim" the AGA/CGA standard, they ought to "live up to it." He suggested that the manufacturers create a "list of their demands which could be discussed point by point" at some time in the near future. Out of these discussions, a committee on "laboratory approvals" was established in June 1951, and the move toward Canadian standards was underway.[19]

The central issue then became which organizations might be accredited to grant the CGA "seal of approval." It was agreed that any such laboratory must first meet some minimum requirements, including adequate facilities and expertise, acceptability to all government bodies, and complete independence from any domestic or international manufacturing interests. Two organizations were approached as possible candidates in 1955: the Ontario Research Foundation (ORF), an independent research corporation with funding from the CMA and the Ontario government; and the Canadian Standards Association (CSA), a non-profit, membership-based standards organization established to represent the interests of business, government, and consumers.

An agreement with the ORF was quickly concluded, but consultations with the CSA were exceedingly difficult. The CSA would not allow for the appointment of more than one CGA member

12: As promotions and advertising became a central part of the CGA's mandate in the 1950s, along with a greater emphasis on recruitment, testing, and policy, novel marketing strategies were devised to give maximum exposure to gas products and services. The all-gas kitchen on the set of Otto Preminger's controversial 1955 film *The Man With the Golden Arm*, starring Frank Sinatra and Kim Novak, was no accident; it was an early example of product placement by the American Gas Association. *Canadian Gas Journal*

to its various standards committees, including those of the Installation Specific Code committee and the Appliance Code Specification committee. No appliances that the CSA tested could bear any "seal of approval" other than its own. The CSA was also mainly concerned with safety, and "lukewarm on the idea of performance tests." What's more, the CSA estimated that the costs of a gas-testing facility would be approximately $70,000, an amount that exceeded the CGA's entire annual budget. The CGA, therefore, decided that although it would collaborate with the CSA on the testing of the electrical components in gas appliances, it would look elsewhere for its gas-related testing needs.[20]

When the CGA's approvals program was launched in 1957, its laboratories included the AGA's Cleveland and Los Angeles testing facilities, along with those maintained by the ORF and the British Columbia Research Council (BCRC), an organization similar to the ORF. In addition to the usual AGA facilities, this

provided Canadian manufacturers with two regional Canadian laboratories to carry out their appliance testing. Any savings from the use of the Canadian facilities would come from the transportation costs, since both of these laboratories decided to charge the going AGA rate. Most provinces were prepared to accept the CGA standards as their own, and Ontario and BC went so far as to make them mandatory.[21]

In a related development, the CGA began to promote better training programs. As the industry expanded and became more complex, more and more workers for specific areas of the gas business were needed. Throughout Canada, skilled labour was already at a premium because of the vibrant economy of the late 1940s and early 1950s. In the short run, some skilled labourers could be lured away from other industries or countries; in the longer term, it was better to train more young people. The CGA contacted officials at the Ryerson Institute in Toronto to help to create a course of study in "Gas Technology." These discussions led to the development of a three-year diploma program embracing basic subjects of the arts and sciences as well as specialized subjects such as "Gas Utilization," "Gas Flow Characteristics," and "Gas Law and Codes." The first class graduated in 1960.[22]

13: When the sophisticated flare-lighting mechanism at a Nevis, Alta., gas-conservation plant failed, a method used since antiquity was adapted to the cause: an archer with a flaming arrow. *Canadian Gas Journal*

Other major initiatives were undertaken in the areas of membership recruitment and general promotions. In June 1954, a special committee on gas-industry development was established to "survey the industry's need for a development program prior to gas being transported across the country, and to advise the executive if and how the CGA can assist in organizing, promoting, and administering such a program." One year later, the committee reported on the need to "(1) bring top-flight executives of member and non-member companies closer to the Association, and (2) securing promises of increased financial support for the Association." Both, it noted, could be achieved by expanding the membership. To assist in doing so, the CGA board of directors approved the appropriation of around $3,000 per year in 1956 and 1957 to be expended at the discretion of the public-relations committee. These monies were then used for the distribution of additional CGA newsletters, reports, and other communications.[23]

All of these innovations paid excellent dividends. During the late 1950s, the association grew to more than 500 members and its revenues increased dramatically. The heavy demands placed upon the association during the later 1940s and early 1950s, a time when its resources were slim, could easily have caused it to drift into obscurity. The fact that it persisted was a testament to the membership's faith in their organization.[24]

1961

1982

Challenges

Previous: The Centennial Flame at the foot of
Parliament was first lit by Prime Minister Lester
B. Pearson to mark Canada's 100th anniversary
in 1967. Fuelled by natural gas, the symbolic
flame – pictured in 2004 – has burned virtually
non-stop for forty years. *David Coombs*

1: One week before Canada celebrated its
one-hundredth birthday in Montreal, the
CGA held its annual convention, appropriately
titled The Challenge of Change, at Jasper,
Alta. *CGA archive*

On 1 July 1967, Canada marked its centennial with an effusion of celebration. Millions of Canadians and tourists converged to join in the festivities at events such as the Canada Festival in Ottawa and the Universal and International Exhibition, popularly known as Expo 67, in Montreal. In 1867 Canada was a rural, agrarian outpost of the British Empire with a population of roughly 3 million people, covering a small band of territory running from the east coast of Nova Scotia, down through the St. Lawrence Valley, and up along the northern shores of the Great Lakes. By 1967 it was an urban, industrial, independent nation with a population of more than 18 million people; it covered a vast transcontinental territory that stretched – as its motto proclaimed – "from sea to sea." Canada had experienced numerous booms and busts, participated in two world wars, and had gone through countless political, economic, and social changes. Its future was to be one of "endless progress" guaranteed by "boundless resources."

In the coming decades, Canada would continue to know progress as well as challenge. Gross National Product increased by an average of nearly 4 percent per year from 1962 to 1973, before slowing to about 1-percent growth per year from 1974 to 1982. The slowing of economic growth, combined with concerns over Canada-US relations, heightening inter-regional tensions, fluctuating economic conditions, and escalating levels of government debt, contributed to a growing sense of unease. Was Canada becoming too tied to its continental partner? Would it be sundered by its internal divisions? Had it lost its competitive advantages? Could it get them back?[1]

The energy sector was regarded as a source of and a solution to these nagging questions. Perceived scarcities led to greater concerns over national energy security and long-term supply. Large-scale public investments brought new competitors into the field. Windfall profits resulting from sudden spikes in prices renewed struggles over resource ownership and taxation. And state-subsidized energy megaprojects were defended as part of Canada's strategy for national revitalization, or opposed as abhorrent examples of economically misguided and environmentally ruinous overdevelopment.

As the CGA learned to navigate these shifting currents it became an influential guide for the mapping of national and regional energy policies. No longer was the association regarded as just a social club for gas executives, or a mere list of contacts that might be consulted for assistance in the case of wartime emergencies or other crises. The CGA, rather, was increasingly recognized as the modern "business organization" that had been envisioned by association president D.K. Yorath in the mid-1950s, one that, under the direction of the first managing director, W.H. Dalton, was equipped with the skills and resources it needed to "assist, advise, and act" on behalf of the gas industry. Between the early 1960s and the early 1980s, the association's membership, assets, and staff expanded well beyond the expectations of its founders, while its consultations with government departments and agencies became progressively embedded within the political process. That is not to say the association always got its

2: In order for the CGA to advocate effectively on behalf of the gas industry, it needed a sound understanding of prevailing attitudes among customers – and non-customers, for that matter. This is the first of a two-part opinion survey conducted in the early 1960s and published in the Canadian Gas Journal. *Canadian Gas Journal*

SURVEY

What does John Q. Public think about natural gas?

This is the first of a two-part series devoted to public opinions on natural gas.

By N. Ivan Wilson

"my local gas utility is owned by the Provincial government—or is it municipal?

"natural gas rates are set by the local gas utility."

"natural gas has no odour and we can't smell it."

"can't grow plants in the house with gas."

These are some of the answers received by the Canadian Gas Journal in a nation-wide poll of Canadians not now using natural gas but living in areas served by gas. The survey covered the provinces of British Columbia, Alberta, Saskatchewan, Manitoba, Ontario and Quebec.

2

way – nor could it, given the multiplicity of interests and imperatives involved in the formulation of government policies. But what the CGA did do was bring the gas industry's perspective to bear on public debates over energy, nudging this discourse in a direction that was more hospitable to CGA members, to government, and to the Canadian people.

Competition in an age of abundance:
The advent of nuclear power and the economics of energy

When the first Canadian Deuterium Uranium, or CANDU, nuclear reactor came on line on 11 April 1962, it initiated a major competitive challenge for the national energy sector. Like hydroelectricity in the first decades of the twentieth century, nuclear power was praised by its promoters in the late 1940s and early 1950s as a "modern," "clean," and "efficient" form of energy for the coming decades. By the late 1950s, many parts of Western Canada and southern Ontario had reached the limits of their hydroelectric generating capacity. Yet as the population and industries of these areas expanded, the demand for power continued to escalate. Though there was potential for further hydroelectric development in Quebec, the Maritimes, and elsewhere, the transmission technology of the time was not yet capable of bringing these supplies to where they were most needed.

This meant the additional demand was to be serviced by a combination of options: coal, oil, and gas. The creation of a nuclear industry thus had tactical and strategic implications for the entire sector. On the ground level, nuclear power was projected to lower the price of electricity in local markets that were underserved by hydroelectric power; at a higher level, it was to stretch the overall capacity of the electrical industry to serve the mass markets of Central Canada and the United States.[2]

Since the early 1940s, the Canadian government had been extremely active in the research and development of nuclear technology. Although its nuclear program was initially connected with British and American efforts, Canada later pressed on with its own independent program after the end of the Second World War. A legislative framework for these purposes was defined in the Atomic Energy Control Act of 1946, which established the Atomic Energy Control Board (AECB) to supervise the "development, application and use of atomic energy." The AECB oversaw this work until July 1952, when the management of the government's Chalk River Nuclear Laboratories was transferred to a Crown corporation, Atomic Energy of Canada Limited (AECL).

The purpose of AECL was to "administer" the Chalk River lab "on behalf of the AECB." Its founding marked the beginning of Canadian research on nuclear power plants, as opposed to reactors intended only for research purposes. In late 1954, AECL called for tenders from private contractors to "design, construct, and supply" a nuclear demonstration plant at Chalk River with the assistance of AECL and the participation of an electrical utility. One month later, Canadian General Electric was chosen as the private contractor and Ontario Hydro was adopted as the utility.

Ontario Hydro's selection came as little surprise. For economic and political reasons, Hydro-Québec preferred to concentrate on hydroelectric power. There was plenty of low-cost hydroelectric potential still to be developed in its waterways, and the provincial government was in no hurry to expand an industry that would be subject to considerable federal oversight. As for

3, 4, 5: With the advent of nuclear power just a few years away, keeping gas in the spotlight was an imperative the industry could not afford to ignore. On 6 September 1963, the inaugural Gas Week was opened in Montreal – by Miss Gas Genie, no less, who is pictured on the previous page lighting the symbolic gas-fuelled torch. The week-long event was a concentrated marketing effort, which combined public lectures, tours, fairs, and tie-ins with appliance dealers and showrooms. Below left, Quebec Natural Gas Corp. president Glenn O. Maddock delivers the opening address; below right, engineers and architects tour the gas-heated Aéroport international Dorval de Montréal. *Canadian Gas Journal*

4

5

the Maritimes and the West, neither region had the size or density needed for supporting large-scale nuclear development, nor did they possess the human and financial resources that could be offered in Central Canada. Moreover, with its hydroelectric capacity and gas-delivery systems stretched close to their limits, Ontario was by far the most obvious big market for nuclear power.

The nuclear demonstration plant was the first to incorporate all of the essential elements of Canada's patented CANDU system: the use of natural (rather than enriched) uranium fuel; the employment of heavy water (a natural substance) as a moderator of the nuclear reaction; and the ability to refuel the reactor while it is in operation (thus reducing the need for shutdowns). All of this made for a plant that could be operated with a high degree of efficiency. Before the demonstration plant was even completed, in 1962, plans were well underway for a full-scale unit at Douglas Point, Ontario. Before the Douglas Point reactor came on line, in November 1966, four even-larger units were planned for Pickering, Ontario. Other provinces soon joined the "nuclear club." Hydro-Québec did so in the early 1970s, with the completion of its first reactor at Bencanour; New Brunswick went nuclear in the early 1980s, with the opening of a reactor at Point Lepreau.[3]

6

6: Propane was put on its mettle at the 1963 Grey Cup, when officials from BC's Empire Stadium called on Vancouver propane supplier Rockgas to dry out the field after the playoff-abused pitch was soaked by a five-day deluge. Industrial heaters had the field ready for the Hamilton Tiger Cats to beat the BC Lions 21-10. *Canadian Gas Journal*

Each of these initiatives reflected the federal government's determination to tap into what the Royal Commission on Canada's Economic Prospects (1957) had called "the promise" of nuclear energy. Investment in nuclear technology was known to have important research applications in energy, medicine, and other areas, and held out the possibility of creating thousands of well-paying jobs. By the late 1970s, there were more than 30,000 Canadians employed in some aspect of the nuclear industry.

But the costs of the nuclear industry turned out to be far higher than anticipated. Hydro-Québec's second power plant, near Gentilly, completed in the early 1980s, was projected to cost $300 million. The total was closer to $1.5 billion. Ontario Hydro's Darlington Station, on which construction began around the time Gentilly was nearing completion, cost roughly double its early estimates of $6.8 billion. Then there was the problem of excess capacity. Throughout much of the 1980s, Hydro-Québec's Gentilly reactor sold power to the United States at below cost in an effort to recoup at least some of its investment, while Ontario's nuclear power plants frequently operated at about half their capacity.[4]

Public investments in nuclear technology came at a time when the demand for power was rapidly escalating. The energy study of the Royal Commission on Canada's Economic Prospects had projected that the nation's power consumption would quadruple between 1955 and 1980. In fact, consumption approximately tripled in that period. The difference was partly the result of conservation measures introduced through the 1970s, which few had foreseen in the 1950s and 1960s. This, however, was still a substantial increase over a relatively short period of time, far more than could be met by Canada's evolving nuclear infrastructure. That being so, established sources of energy were needed to bridge the divide between the "promise" and the "reality" of new technology.[5]

Natural gas played a huge role in this regard. The 1960s witnessed the largest additions to gas reserves in the history of the industry up to that point. In 1964 alone, gas reserves increased by 17.4 percent, bringing Canada's total proven reserves to 43.4 trillion cubic feet (or 31 years of supply at the 1964 rate of production). The total mileage of gas pipelines, which had once consisted of a few hundred miles organized around a handful of regional hubs, stood at 41,000 miles spanning the entire country, with spur lines into the massive US distribution system.[6]

With this solid foundation, gas services came to supply an ever-larger segment of national energy needs. Natural gas furnished

Table 7.1 Canadian energy consumption by primary source, 1960-1980 (percentage distribution)

	Year		
Source	1960	1970	1980
Petroleum	56.5	53.3	49.1
Natural gas	12.9	23.4	26.0
Coal and coke	18.8	13.3	12.6
Hydroelectricity	11.8	10.0	10.6
Nuclear electricity	–	0.1	1.7
Total	100	100	100

Source: CGA, *Canadian Gas Facts* (1981), 3.

Table 7.2: Natural gas utility sales by class and province, 1960-1980

Province/Year	Vol. of sales (million cu. ft.)	Class of service (percentage distribution) Residential	Commercial	Industrial	Total
New Brunswick					
1960	81	98.8	1.2	–	100
1970	62	53.2	46.8	–	100
1980	77	21.7	78.3	–	100
Quebec					
1960	12,177	35.4	11.3	53.3	100
1970	50,705	28.8	13.4	57.8	100
1980	103,120	18.2	17.1	64.7	100
Ontario					
1960	103,864	40.5	12.9	46.6	100
1970	406,000	24.9	19.9	55.1	100
1980	658,062	21.6	25.6	52.8	100
Manitoba					
1960	11,535	39.7	20.7	39.7	100
1970	51,528	38.9	30.0	31.1	100
1980	62,931	35.6	36.6	27.8	100
Saskatchewan					
1960	30,433	34.0	15.6	50.4	100
1970	79,660	28.0	15.8	56.2	100
1980	97,485	28.9	22.1	49.0	100
Alberta					
1960	136,824	26.6	17.9	55.5	100
1970	232,700	24.5	22.3	53.2	100
1980	452,530	19.1	19.1	61.8	100
British Columbia					
1960	25,788	40.6	14.2	45.2	100
1970	96,786	27.5	19.2	53.3	100
1980	152,757	27.2	26.2	46.6	100
Canada					
1960	320,701	33.8	15.6	50.6	100
1970	917,441	26.4	20.3	53.4	100
1980	1,526,961	22.2	23.4	54.4	100

Source: CGA, *Canadian Gas Facts* (1981), 8 and 12

12.9 percent of the national energy consumption in 1960; by 1980, it furnished 26 percent. Hydroelectricity, by contrast, delivered 11.8 percent of energy supplies in 1960 and 10.6 percent in 1980, while nuclear power provided nothing in 1960 and just 1.7 percent in 1980 [Table 7.1]. Though there were substantial variations in the "energy mix" of different parts of the country, natural gas remained an important part of the blend in almost every branch of service throughout each region of Canada [Table 7.2].

The postwar years were a time of transition for gas services. From being the preferred choice for home cooking, gas ranges were becoming a distant second to their electric counterparts. In his presidential address to the CGA's 1965 annual convention, J.W. Kerr spelled out the troubling statistics. "Since 1949," he explained, "the number of households in Canada has increased 41.6 percent. In the same fifteen years, electric range shipments increased 207 percent and gas ranges by only 26 percent. In 1963 ... electric outsold gas ranges [by a ratio of] 6.25 to 1 – 277,500 to 44,500." Kerr blamed multi-million dollar electric-appliance advertising campaigns, but continued additions to electrical generating capacity could not have helped. Nor could the gradually shrinking interest on the part of gas utilities, for which the residential market was becoming a less and less important outlet for their product. Between 1960 and 1980, residential customers plunged to 22.2 percent from 33.8 percent of total gas sales in Canada.[7]

Declines in domestic cooking were offset by gains in home and commercial heating, as well as those in the expanding commercial and industrial applications of gas services. From 1960 to 1975, for instance, factory shipments of central-heating, warm-air furnaces increased to 118,100 units from 65,800; shipments of automatic water-storage heaters swelled to 159,300 units from 108,500; and those of automatic clothes dryers jumped to 10,100 units from 8,200. Taken together, these changes formed part of the industry's overall shift toward commercial and industrial gas sales.

AS EUROPE AND ASIA WERE PIECING THEMSELVES BACK TOGETHER, CANADA WAS SOARING TO NEW HEIGHTS WITH THE HELP OF ITS WARTIME ADVANCES IN PRODUCTIVE CAPACITY AND TECHNOLOGICAL EXPERTISE

Exports experienced even more robust growth. In 1960, domestic sales of natural gas were valued at $194 million and exports at $22.3 million; by 1980, domestic sales were worth more than $3.6 billion and exports $4 billion. This made exports, virtually all of which were destined to the United States, just as important as domestic sales. Most of the time this was not much of an issue. Canadians and Americans shared many of the same values and world views, as well as a long history of economic and cultural exchange. Moreover, the Canada-US trade in natural resources was a big part of the positive side in Canada's overall trade balance, bringing in US dollars that helped promote economic development and finance trade in other areas. That said, as with any intense bilateral relationship, there were bound to be strains along the way.[8]

The ambiguous milieu of energy policy:
Liberalism, nationalism, and environmentalism

The development of nuclear power, the extension of social security, and the building of the TransCanada pipeline were manifestations of Canada's postwar efforts to create a stronger and more equitable society. Yet Canada still faced numerous challenges. More and more, there was a sense that other countries continued to exert too much influence on Canadian affairs, that government had reached the limits of what it could afford to do for its citizens, and that economic progress had gone too far. Throughout the 1960s and 1970s, this complex intermingling of confidence and insecurity made for a volatile mixture.

By this time, Canada had moved well beyond its colonial dependence on Great Britain. Canada had been Britain's lifeline through the two world wars and had provided it with generous assistance through the mid-1940s. As the countries of Europe and Asia were piecing themselves back together after the end of the Second World War, Canada was soaring to new heights with the help of its wartime advances in productive capacity and technological expertise. It now possessed not

7: As governments expanded the safety net of social assistance in the 1960s and 1970s, subsidized housing projects, renewals, and retrofits became another lucrative source of business for gas companies throughout the country, such as Union Gas in southern Ontario. *Union Gas archive*

only vast natural resources, but a vibrant and competitive manufacturing sector, a skilled workforce, and a growing population fuelled by immigration.[9]

The nation's new-found wealth allowed for expanded social programs. A universal old-age pension plan was established in 1951, replacing the "means-tested" system that had been around since the late 1920s. It was later widened into the Canada Pension Plan and Québec Pension Plan in 1965, with their higher standards. Hospital insurance came in 1957, supplemented by insurance covering physician services in 1966. Unemployment insurance, which policymakers had long been hesitant to implement in the first place, was expanded in the 1960s and 1970s.

Upon this layer of individual assistance, an elaborate system of subsidies to "underdeveloped" regions of the country was overlaid. The federal government had been active in attempting to foster regional development for some time with the Prairie Farm Rehabilitation Act (1935) and the Maritime Marshlands Rehabilitation Act (1948), to which it now added the Agricultural Rural Development Act (1961), the Atlantic Development Board (1962), and the Fund for Rural Economic Development (1966). Toward the end of the 1960s, these programs were tied together under an entire department – the Department of Regional Economic Expansion.[10]

Canada also took up a more prominent place in world affairs. Canadians made real contributions to the framing of postwar international institutions and agreements such as the International Monetary Fund (1945), the General Agreement on Tariffs and Trade (1947), and the North Atlantic Treaty Organization (1949). The federal government extended foreign aid to developing countries through programs such as the Columbo Plan (1951), through which it lent assistance throughout Southeast Asia. Elsewhere, the military engaged in peacekeeping operations to improve international stability in faraway places such as Egypt and Cyprus, nurturing the notion that Canadians had an international mission that won respect abroad, and adding to self-esteem at home.[11]

Despite these achievements, the nation seemed vulnerable on several fronts. Early signs of trouble appeared with the first postwar recession, in the late 1950s. Recovery returned in the early 1960s, but inflation and unemployment continued to creep upward over the next two decades. According to contemporary economic textbooks, the latter was supposed to be impossible. Yet there it was. By the late 1970s, unemployment hovered at more than 7 percent of the labour force, and prices rose about 8 percent each year. Tougher economic conditions, population growth, bigger social programs, and tax changes fed the growth of government debt, too. In the late 1960s, the federal government owed $18 billion and interest payments constituted 12 percent of its budget; in the early 1980s, the debt reached $206 billion and interest payments 20 percent.[12]

At the same time, there was concern that Canada's former dependence on Britain was being replaced by an ever-greater reliance on the United States. During the darkest moments of the Second World War, following the fall of France, Canada entered an intimate military and economic alliance with the United States. After the end of the war, Canada found itself awkwardly sandwiched between the world's two new superpowers, the United States and the Union of Soviet Socialist Republics. Canada aligned itself with its next-door neighbour and closest friend, steeling itself for the coming Cold War. A system of Distant Early Warning (DEW) radar stations were constructed in the Far North in the early 1950s, and defence integration was taken a step further with the signing of the North American Aerospace Defense Command (NORAD) agreement in 1958. Under NORAD, continental air defence was placed under the supreme command of an American officer, with a Canadian officer as the deputy commander. For some, this made Canada a "junior partner" in the defence of North America; for others, it was merely recognition of geopolitical realities. Both the United States and the USSR possessed nuclear weapons, and neither was likely to ask for permission to use Canada's airspace in any prospective conflagration.[13]

Canada was also becoming more economically tied to the United States. In the years leading up to the First World War, Britain supplied almost 80 percent of Canada's foreign investment and accounted for roughly half of its external trade. Forty years later, these same figures applied to Canada's economic relations with the United States. Canada-US integration only deepened from there, with the United States coming to account for as much as 60 percent of Canada's trade toward the mid-1970s.[14]

Geography, economics, and history also fostered close cultural connections. Here, too, the United States was the dominant partner, with nine times the population and sixteen times the Gross National Product of Canada at mid-century. Canadians happily tuned in to American music and television programs, and bought up US magazines and other products. At least a few

Canadian entertainers, such as William Shatner, Joni Mitchell, Robbie Robertson, and Anne Murray, had success south of the border as well. At Canadian universities, meanwhile, nearly half of the faculty was imported from the United States in the late 1960s. Conversely, thousands of Canadians were teaching at US universities or working in other US industries. The border hardly existed.[15]

But some thought it must. Reporting in 1951, the Royal Commission on National Development in the Arts, Letters, and Sciences concluded that Canada's culture was derivative and underdeveloped. Later that decade, the Royal Commission on Canada's Economic Prospects had become the first of a host of government investigations to sound the alarm over foreign investment in Canada. Foreign-owned firms, it was said, paid less in taxes, created fewer jobs, and conducted less research and development than companies owned by Canadians. Criticism of US influence on Canada, which drew from and built on these notions, became fashionable for a time.[16]

Policy responses were soon forthcoming. Federal funding for universities and cultural initiatives were dramatically increased, and in the early 1970s a federal Foreign Investment Review Agency (FIRA) was established to watch over new investments, while Canadian trade officers were instructed to strengthen the nation's economic ties with Europe, Asia, and Africa. The effects of these measures were mixed. Canadian culture undoubtedly benefited from the new financial support, but American cultural influence was hardly diminished. Throughout its lifetime from 1974 to 1985, FIRA approved more than 80 percent of the cases it reviewed. And in the years when diversification was most vigorously pursued, Canada's trade with the United States became more important than ever before. Symbolic or not, these measures reflected mounting apprehensions over the sometimes stifling continental embrace, as well as the idea that government ought to do something about it.[17]

victimizing vegetation?

8: By the early 1970s, the environment was emerging as a complex challenge for Canada's resource-based economy. This detail from a Union Carbide advertisement, circa 1968, makes the point very bluntly that crop and habitat destruction through gas leaks is 'business suicide'. *Union Carbide/Canadian Gas Journal*

(FPLL), comprised of the Alberta Gas Trunk Line Company and Westcoast Transmission, called for a shorter route, down the Alaska Highway to Alberta.

Early on in the process, in 1974, the federal government appointed a royal commission led by Justice Thomas Berger to assess the social and environmental implications of the proposed development. Many native and environmental groups made representations to the commission, expressing their dismay. In the end, the Berger report, issued in 1977, rejected any construction in the northern Yukon and proposed a ten-year moratorium to settle land claims and allow time for northern aboriginals to prepare for the changes that the development would bring to

IN 1974, THE FEDERAL GOVERNMENT APPOINTED A ROYAL COMMISSION TO ASSESS THE SOCIAL AND ENVIRONMENTAL IMPLICATIONS OF THE PROPOSED ROUTE THAT WOULD RUN ACROSS THE NORTHERN YUKON AND DOWN INTO ALBERTA

The environment provided an additional challenge. For most of their nation's history, Canadians had been primarily preoccupied with the task of seeking to carve out a decent living from a harsh landscape and an unforgiving economic climate. As urban and industrial development proceeded, however, there was rising concern about the long-term costs of progress. In the mid-1970s, the seriousness of these concerns became readily apparent in the debates over the plans to construct an Arctic gas pipeline from reserves in the Beaufort Delta to the markets in Canada and the United States. There were two main proposals. Canadian Arctic Gas Pipeline Limited (CAGPL), a consortium of twenty-seven Canadian and American oil and gas companies, planned a route that would run across the northern Yukon and down the Mackenzie Valley into Alberta. Foothills Pipe Lines Limited

their communities. Subsequently, the National Energy Board opted for the FPLL proposal, partly on the basis of its higher Canadian content in ownership and partly because of the less disruptive social and environmental effects of the Alaska Highway route. Though changing conditions forestalled these plans as well, the Arctic-pipeline experience was but another portent of the gas industry's changing policy universe.[18]

Predictions of the never-ending growth of energy demand, faith in the power of government action to resolve economic and social problems, fears of American domination, and concerns over the long-term effects of urban-industrial development: these were the broad-stroke issues that defined the energy policy of the day.

9, 10: Infrastructure had become highly sophisticated by the 1970s – perhaps just as sophisticated as the political and economic milieu of the same period. The top photograph shows a mechanical control room in the 1920s; the bottom is a computerized Union Gas control room in the mid-1970s. *Canadian Meter Company, Union Gas archive*

9

10

The 'real world' of energy politics:
From the oil shocks to the National Energy Program

The late 1950s and the 1960s were relatively stable for the gas industry; the 1970s and 1980s, however, were anything but. Price instability and "excess profits" brought increased government intervention, which culminated in the federal government's abortive effort to create the National Energy Program (NEP), aimed at "Canadianizing" the oil and gas industry, capturing a larger federal share of oil and gas revenues, and controlling domestic prices. It was a trying and uncertain time for gas companies, governments, and consumers.

The domestic energy market received its first international shock in the aftermath of the Yom Kippur War. On 6 October 1973, Egypt and Syria launched simultaneous attacks on Israel in an attempt to regain territory that had been lost in previous conflicts. Although initially successful, the invading forces were soon turned back, on the strength of Israel's US-supplied weapons technology and military intelligence. On 24 October, the United Nations brokered a ceasefire, and peacekeeping troops, including Canadians, were dispatched to the region. This, however, was far from the end of the conflict. The territorial issues were unresolved, and much of the Arab world was furious over the American support of Israel. In retaliation, Saudi Arabia, Libya, Dubai, Qatar, Bahrain, and Kuwait cut off all oil supplies to the United States. Several of these countries were members of OPEC, the Organization of Petroleum Exporting Countries, formed in 1960. Securing higher oil prices had always been among OPEC's goals, but it had never sought to do so by means of such drastic reductions in supply. This changed after the Yom Kippur War, and an intensified resoluteness developed among OPEC members to earn more from oil sales. Within months, the price of oil rose fourfold, from about $3 to $12 per barrel. Today, that would be equivalent to a jump to $40 from $10.[19]

Oil prices stabilized at this higher level for much of the 1970s, until a second shock in 1979. Early that year, a political revolution in Iran replaced Shah Mohammad Reza Pahlavi with Ayatollah Ruhollah Khomeini. The disruption of Iran's contributions to the international oil supply was among the immediate effects of the turmoil that followed. Prices shot upward again, this time going from $15 to $40 per barrel – or the equivalent of $80 in today's dollars. Some mainstream oil forecasters mused that prices were bound to rise much further. Indeed, a large portion of the short-term price increases at the time were driven by wildly inaccurate assessments of the market's ability to respond to changed conditions. Forecasters miscalculated the industry's potential to increase production, OPEC's ability to maintain unity, and the consumer's capacity to find alternatives and curtail demand. Certain government interventions also exacerbated matters by distorting the market's price adjustments, though public-opinion surveys indicated that large numbers of consumers suspected that the machinations of multi-national corporations were at work.[20]

A shift in energy policy was already underway. In 1971, the federal government had appointed an inquiry to review the entire range of energy-related demand, supply, ownership, and regulatory issues. Because it reported in June 1973, its long-term economic assumptions were quickly dated by world events; however, its short-term political impact was to raise the possibility of a state-owned oil and gas enterprise. The latter was among the main demands of the New Democratic Party, which at that time held the balance of power in a minority parliament. Its support was crucial to the survival of the ruling Liberals. In December of that year, the federal government announced its intention to strengthen the "Canadian content" of the oil and gas sectors and to provide policymakers with a better window on the key areas of supply,

BETWEEN 1974 AND 1979, THE DOMESTIC PRICE OF GAS WENT FROM $0.62 TO $2.15 PER 1,000 CUBIC FEET AND THE EXPORT PRICE WENT FROM $1 TO $2

costs, and pricing. A new company, Petro-Canada, came into being on 30 July 1975 and rapidly expanded its presence. After starting from nothing in the middle of the decade, by the end of the 1970s it had become Canada's "sixth-largest producer of oil and gas, with assets in excess of $4 billion." That the Liberals proceeded with Petro-Canada even after regaining a majority in 1974 was indicative of the popularity of the company's goals and the policy momentum toward greater state involvement in the oil and gas sector that had gathered at the federal level in the period between the late 1950s and the mid-1970s.[21]

Provincial governments were equally determined to exert their influence. Since the end of the Second World War, the provinces had been progressively expanding their responsibilities, budgets, and bureaucracies. For the first time, many had developed the policy capacity to meet the industry and the federal government on a more equal footing. To assist in doing so, to promote development, and to secure a larger share of resource revenues,

Saskatchewan founded Sask-Oil in 1973 and Alberta created the Alberta Energy Company (AEC) in 1975. Sask-Oil was a 100-percent publicly owned Crown corporation; AEC was a 50-50 public-private venture.

As the state expanded its presence, a mixture of factors placed greater emphasis on conservation. In November 1971, the NEB denied three separate applications to increase gas exports to the United States, concluding that Canada's existing reserves were insufficient for its future needs. It was the first time the NEB denied all export applications, though in August 1970 it had denied one application and reduced the terms of three others. After the first oil shocks of 1973-74, the regulated price of gas in Canada was increased in several stages to reflect the higher value of energy products and to provide larger revenues for exploration and taxes. Between 1974 and 1979, the domestic price of gas went from $0.62 to $2.15 per 1,000 cubic feet and the export price went from $1 to $2. Higher prices generated greater interest in reducing consumption, as did the appearance of scientific studies, such as those of the Club of Rome, which argued that the world was approaching "the limits of growth."[22]

Government and industry sought to provide assistance. At the federal level, Environment Canada was formed in 1970, and the Office of Energy Conservation was created within the Department of Energy, Mines, and Resources in 1974. Alberta followed suit with the first provincial environment ministry in 1971. The CGA founded the Canadian Gas Research Institute (CGRI) in 1974 to "foster the more effective use of materials and resources" and to develop "new equipment and appliances for the ultimate benefit of consumers." The institute was funded by eighteen sponsoring member companies along with a grant from the federal Department of Industry, Trade, and Commerce. Not long after its appearance, the CGRI was invited to collaborate on similar research conducted by the US Gas Appliance Manufacturers Association, a "reciprocal exchange" that the CGA board of directors regarded as "invaluable."[23]

11: With the oil shocks of the 1970s spiking prices at the pumps, alternative fuels for vehicles became an attractive option. This led to the development of passenger vehicles that burned propane or natural gas, such as this prototype sponsored by Union Gas, circa 1984. *Union Gas archive*

11

During the early 1980s, the apex of government intervention came with the NEP. As academics Bruce Doern and Glen Toner point out, the elements that gave rise to the policy had been circulating for years in the form of discontent over energy supply, foreign ownership, taxation, and prices. As early as April 1976, the Department of Energy, Mines, and Resources released a nine-point "Energy Strategy." It called for more vigorous federal efforts to ensure national energy security, increase Canadian ownership and participation, control pricing, and promote research, development, and exploration. Many of these themes were echoed in Liberal leader Pierre Trudeau's "Halifax statement" on energy policy during the 1980 federal election campaign. During their brief term of office in 1979-80, the Conservatives, who had been vociferous opponents of Petro-Canada, had also proposed to move toward energy self-sufficiency, Canadianization (through increased "private opportunities" for Canadians), and federal-provincial revenue redistribution. The Tories' only mistake was to seek parliamentary approval of higher gas taxes when they did not control the House of Commons. Their first budget was defeated, and the Liberals were returned to power with a majority government in the ensuing election.

12: The Northern Ontario Natural Gas Co.
and Orenda Engines collaborated on a
gas-fired turbine that was selected to
power RCAF Station North Bay. The
base, later renamed CFB North Bay,
was a command centre for NORAD
ground-to-air Bomarc missiles in the
Cold War. *Canadian Gas Journal*

12

Self-sufficiency, nationalization, revenue redistribution, and fair pricing became the main principles of the Liberals' energy policy. Under the terms of the NEP, first announced in October 1980, more than $2.5 billion would be made available for energy conservation, renewable-resource development, and oil-substitution initiatives. Another $2.5 billion would support exploration and development on the "Canada Lands" in the Far North and offshore, where the provinces had no jurisdiction. These grants, however, would only be available to those firms that were at least 50-percent Canadian-owned and -controlled. Furthermore, the federal government reserved the right to take a 25-percent "back in" position on all new and existing leases on the same lands. There would also be more money for research and development, Petro-Canada, and various other initiatives. To fund the estimated $8-billion cost of the program, Ottawa would increase its share of total oil and gas revenue to 24 percent from 8 percent between 1980 and 1983, while the provinces and industry would see their shares reduced to 31 percent from 45 percent and 45 percent from 47 percent, respectively. Consumers would pay more for oil and gas as a result of increased taxes, but could still expect to pay less than the projected world prices.

Initial reaction to the NEP varied widely. Multi-national companies were "shocked and angered" by the Canadianization initiatives, and the Alberta provincial government was even more stunned and enraged over the dramatic cut in its share of total revenues. Alternatively, opinion polls suggested many Canadians supported the NEP, with as many as 64 percent favouring "even faster Canadianization." Surveys conducted by the Canadian Petroleum Association (CPA) indicated that part of the problem was the low standing to which the industry had fallen in the public view.[24]

The NEP, therefore, prompted oil and gas industry associations to adopt a more active role in government relations and public debates. The CGA opened its first branch office in Ottawa, and released a policy statement on the program's "impact on the Canadian consumer and the natural gas industry." Although certain aspects of the NEP seemingly worked to the benefit of the gas industry by promoting oil-to-gas conversions, the CGA contended that the program had a negative effect on Canadians and the energy sector as a whole. It noted, for instance, that the new taxes applied to oil and gas tended to distort the relative costs of these sources of energy vis-à-vis "electricity, coal, and various renewable energy forms." Along the same lines, it questioned the logic of having oil and gas consumers pay the full costs for the Canadianization of these industries, if such programs were to serve the national interest. Moreover, it warned that the uncertainties introduced by lags between government intentions and policy in the oil-substitution program had actually delayed the rate of conversions.[25]

13, 14: As a complement to advertising, the CGA began producing and
distributing educational material in the 1970s. The goal was not
simply to sell more gas per se, but to ensure the public was getting
timely, accurate information about the sources and uses of natural
gas, as well as the industry in general. Natural Gas at Work booklet,
circa 1977. *CGA archive*

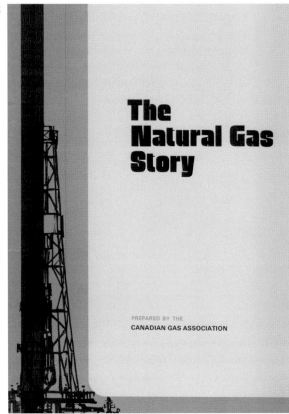

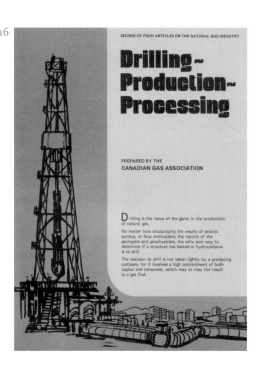

15, 16: Perhaps one of the most successful of the CGA's print initiatives was The Natural Gas Story, a four-part series of booklets that follow the gestation of natural gas from the dinosaur era to its arrival in the home. The suite of booklets was found in homes, schools, and public libraries throughout the country. *CGA archive*

THE PUBLIC-RELATIONS DEPARTMENT CONTINUED TO TURN OUT RELEASES, TV COMMERCIALS, FILMS, SPEAKER KITS, AND PUBLICATIONS

Outsider to insider:
The CGA from the early 1960s to the early 1980s

But it was economics, not politics, that killed the NEP. The entire edifice of the program – exploration, development, Canadianization, and taxation – was based on the assumption that oil would remain above $30 per barrel, as forecasters had expected. Instead, oil prices crashed. Between 1981 and 1985, crude-oil prices plummeted from about $35 to just slightly more than $14 per barrel. Exploration and production collapsed, prompting a progressive redistribution of taxation in an attempt to attract business back into the field, while government debt steadily increased, thanks to the cost of the NEP and the onset of a deep recession. In less than four years, the NEP was in tatters and national energy policy was back to where it had been in the early 1970s.[26]

The impetus toward centralization and nationalization was also evident at the CGA, as the scale and scope of the association's operations broadened more than ever. By the early 1980s, the CGA had more than 641 members, a staff of 150 employees, assets of more than $1 million, and direct involvement in almost every area of gas policy.[27]

The CGA's Gas Measurement School was among the most successful endeavours during this very active period. The program was designed to train gas company personnel in the technical aspects of managing and measuring gas resources. The first two-day event, in 1962, attracted more than 115 technicians, engineers, and sales personnel from thirty-five utilities, transmission companies, and manufacturing companies. By pooling their resources and sharing information, gas companies not only saved on training costs but also gained access to a wider array of expertise. Perhaps more than anything, these advantages explain why the measurement school has carried on its contributions into the twenty-first century.[28]

The CGA offered several other educational opportunities. The gas-technology program at Ryerson continued into the mid-1960s, though enrolment was in decline. At a CGA panel on manpower development in 1964, Don Jones of the Procter and Gamble Company ruefully observed that the program had only two graduates in 1963. Unless one could be sure of obtaining a job in the gas industry, the diploma program was too specific. From the industry's perspective, it was preferable to be able to draw from a larger pool of candidates with varied skills and backgrounds. These factors propelled a shift away from the college-based program toward more CGA-sponsored gas-technology certificate programs and workshops, which could be adapted to meet the needs of the membership's existing personnel.[29]

The CGA, however, did not abandon its recruitment efforts. As Jones saw it, part of the difficulty faced by the industry was that it was perceived to be, like the civil service, "good, safe, necessary, but certainly not very exciting!" The other problem was the lack of "literature on specific opportunities in the gas and oil industry." At the placement offices of one of Canada's largest universities, Jones found only "two thin gas company files with annual reports and facts about the gas industry (including one piece on how a Bunsen burner operates)." Things looked up with the initiation of the CGA's personnel-management division in the 1970s, which, by decade's end, could boast of having produced "seventy-seven technical papers on a variety of aspects of the industry, and a gas industry careers pamphlet that has seen wide distribution."[30]

Yet the heart of the CGA's activities remained in the promotion of gas services. In the early 1960s, the marketing engineer's

department was created to advise "business and government executives, architects, engineers and other design agencies" on the benefits of natural gas. During the first six months of its operation, the department made proposals on as many as twenty-five projects to various government agencies, compiled a mailing list of more than one thousand architects and consulting engineers, and generated a "central file of technical papers, briefs and brochures" for the use of CGA members.[31]

The sales and promotion department augmented the work of the marketing engineer. In the last eight months of 1962, it gave away more than 20,000 of the "gas industry's most popular salesman," the "little Gas Genie," as part of its members' promotions for customers purchasing gas equipment. Three years later, it introduced seventeen TV commercials, featuring themes such as "It's not the same without the flame" and "Natural gas runs circles around every other fuel," each of which was produced at a cost of "one-tenth of a custom-built commercial prepared by an individual company." Reorganized as the public-relations department in the 1970s, the department continued to turn out releases, TV films, speaker kits, and other materials for use by television stations, press, and members. One well-received product was *The Natural Gas Story*, a four-colour booklet widely circulated by CGA members. The most ambitious project was a twenty-six minute feature film on the gas industry in Canada, past, present, and future." Seventeen member companies contributed to the making of the film, sixty-four prints of which were in circulation across Canada in 1977. According to that year's report from the CGA's board of directors, the film was "in high demand from schools, service clubs, the media, and government agencies."[32]

The CGA's laboratory testing facilities were another key aspect of the association's market-building activities. In 1962, the CGA assumed full responsibility for the administration of its safety-testing laboratories at the Don Mills head office. From that time forward, the Ontario Research Foundation would act only as a "technical consultant," carrying out an annual independent

17: The CGA seal of approval was administered to appliance manufacturers after their products were put through rigorous testing at the CGA labs in Don Mills. The page spread pictured is from a brochure titled The Story of the Canadian Gas Association's Seal of Approval, which was distributed to CGA members, appliance manufacturers, and consumer showrooms. *CGA archive*

17

it's a medal of honor every gas appliance hopes to wear

A SEAL OF SATISFACTION FOR YOUR GAS APPLIANCES

When you buy a gas appliance ask to be shown this little seal:

This is the *Approval Seal* of the Canadian Gas Association. An exact duplicate of any appliance which wears it has had to pass merciless and terrific tests in the Association's impartial, non-profit Laboratories in Toronto or Vancouver.

These Laboratories are maintained by the gas industry to give stringent and exhaustive tests to all types of gas appliances; to check on their substantial construction, their safe operation, their dependability and their satisfactory performance.

These Are Your Labs.

These Laboratories exist solely to represent YOU — and they do represent you . . . most effectively. No manufacturer can influence their decisions in any way. They are *your* Laboratories, operated *for you* by the gas industry.

I'd like to try for the C.G.A. Seal of Approval   *You'd better be good*

And they are unique — they are the only testing laboratories operated by an *industry* for its *customers!*

How to Win Hundreds of Thousands of Customers a Year

This hard-earned badge of merit — and the integrity it represents — help a lot to explain the marvelous growth of the gas industry in recent years. It also helps explain why the gas industry is winning new customers every minute.

Do you know that there are now more than 93,000,000 gas appliances in use in North America. Well over 7,000,000 persons in Canada now depend on gas for their everyday home cooking, comfort and convenience, and they regard it as the cleanest, most dependable fuel they can buy.

audit of the CGA facilities to maintain consumer confidence in the program. Three years later, an agreement was concluded with the Canadian Standards Association that would finally see "CSA approval requirements issued under the sponsorship" of the CGA. As the CSA recognized, it was obviously in the CGA's best interests to hold its safety laboratory to the highest possible standard. This allowed one committee of the operating division to oversee the entire process and justified the appointment of a full-time staff engineer as the manager.[33]

The manufacturers' section of the association was essential to the effective operation of the CGA labs. As well as providing important technical data, it kept close contact with all levels of government and helped to keep track of the increasing volume of codes, regulations, and standards churned out by their various agencies. Its many representations to government in the 1970s also included meetings with the federal advisory committee on industrial benefits for natural-resource development, to discuss increased involvement of Canadian manufacturers and the Ontario Energy Branch to resolve "ambiguous clauses" in the Ontario Building Code.[34]

AN AGREEMENT WAS CONCLUDED WITH THE CANADIAN STANDARDS ASSOCIATION THAT WOULD FINALLY SEE 'CSA APPROVAL REQUIREMENTS ISSUED UNDER THE SPONSORSHIP' OF THE CGA

As state intervention at all levels and in all aspects of the gas industry proliferated, the management of government relations became an ever-greater part of the association's main concerns – a changing emphasis signified by the reorganization of the marketing engineer's department into the government and corporate relations division in the mid-1970s. As laid out by the board of directors in 1975, the objectives of the new division were "increased government relations at all levels; liaison in the field of consumer and corporate affairs; long-range planning; energy studies and customs and excise taxes." Many members maintained their own government and business relations departments or agents as well, to which the CGA supplied invaluable support.[35]

Other committees and sections of the association looked at everything from the challenges of metric conversion to the compilation of gas statistics. All told, 77 committees and subcommittees held 106 meetings in 1964-65 alone, and the board of directors met "five times in various cities across Canada while the Executive Committee held another four meetings." Collectively, it was estimated this represented roughly 17,000 hours expended upon the association's affairs by 1,900 CGA member-company employees and associates. Financially, the association was in such shape that it had been able to pay off the mortgage on its head office in just three years. It was an exciting time to be a member of the association or part of its staff. Before long, however, sweeping changes would dramatically re-orient national politics, the industry, and the association.[36]

1983

2007

8

Change

Previous: The election of Margaret Thatcher in Britain, Ronald Reagan in the US, and Brian Mulroney in Canada ushered in a new era of global capitalism in the 1980s. Reagan, Thatcher, Mulroney, and Japanese Prime Minister Yasuhiro Nakasone are shown attending the 1986 G7 summit in Tokyo. *Thatcher collection*

1: The energy sector was targeted by the Trudeau Liberals as one worth investing in. Success in energy, according to the plan, would benefit all Canadians. This meant more money to finance intricate and ambitious projects, such as offshore gas exploration in Atlantic Canada. *CGA archive*

As debate on the National Energy Program peaked in late 1981, the Liberal government of Pierre Trudeau unveiled a major discussion paper outlining its economic strategy for Canada in the 1980s. The strategy stressed the development of Canada's natural resource industries, singling out the energy sector as the lynchpin of the nation's economic and industrial future. The government set out a plan for heavy investments in large-scale energy projects, such as the oil sands of Alberta, the Canada Lands of the Far North, and offshore in Atlantic Canada. These investments were to increase Canadian resource exports, create employment in underdeveloped regions, and raise tax revenues. The surplus of federal income generated was then to be plowed into financing industrial restructuring in general, as well as encouraging manufacturers to supply the "machinery, equipment and material needed for resource development, and to extend the future processing of resource products beyond the primary state" in particular.

This collection of policies represented the culmination of nearly a decade's worth of research and discussion among representatives from government, industry, and academia. It was, as scholars Michael Howlett and M. Ramesh point out, an amalgam torn between "interventionist and free market approaches, nationalist and continentalist approaches, and between proposals that emphasized manufacturing and those that emphasized the resource sector." But with the onset of a severe recession in 1982, and the simultaneous collapse of commodity prices, the paper quickly became "irrelevant as a guide to policy."[1]

The shift to the free market: the Macdonald commission

With Canadian economic policy having reached another crossroads, the federal government established the Royal Commission on the Economic Union and Development Prospects for Canada in late 1982. Donald S. Macdonald, who had been the Liberal minister of energy, mines, and natural resources from 1972 to 1975, chaired the commission. The most extensive reassessment of Canada's economic strategy since the late 1950s, it sat for nearly three years, receiving hundreds of submissions, and generated more than seventy volumes of background research.[2]

The commission's final report recommended a move away from the interventionist and nationalist tendencies that characterized federal economic policy from the early 1970s to the early 1980s. In the face of "rapidly changing international and domestic economic circumstances," it instead urged greater reliance on market mechanisms as the best way to promote economic development. The commission specifically rejected industrial policy based on government attempts to select and develop the "winning" industries of the future, stating it had "little confidence" that centralized governments were better positioned to make such decisions than the decentralized networks of producers in the free market. If nothing else, the experience of the NEP appeared to demonstrate this stark reality quite clearly. The commission further called for the negotiation of a comprehensive, bilateral free-trade agreement with the United States in order to secure access to Canada's major trading partner and increase productivity by exposing the national economy to more international competition. At a time of rising trade protectionism in the United States and falling workplace productivity in Canada, the commission held that this too would be crucial for the nation's economic future.

The Macdonald commission's report reflected the growing preference for free-market responses to economic and social problems throughout western societies in the late 1970s and early 1980s. Although the reasons for this ideological shift are complex, two main factors stand out. The first was the loss of faith in government planning and regulation. At the end of the Second World War, macroeconomic management seemed to hold the key to resolving the main economic and social problems: balancing the business cycle, maintaining the momentum of economic expansion, and reducing the scourge of unemployment and poverty. By the late 1970s, however, rising government debt and declining economic performance began calling the existing approach into question. At the same time, new policy challenges were appearing on the political agenda. During the immediate postwar period, economic stability and full employment had been the main issues. But three decades later these would be joined by growing concerns about efficiency and inflation, problems for which government planning and regulation seemed to have few solutions.

Such developments promoted the formation of very different policies. Broadly speaking, these policies shared two related assumptions: that reducing external barriers to trade and investment was desirable, and that this should be coupled with greater use of market mechanisms in the domestic economy. In the midst of the economic troubles of the late 1970s and early 1980s – while "economic nationalism" was in vogue and western economies were in the doldrums of recession, inflation, and unemployment – this free-market alternative provided a powerful criticism of the existing paradigm, along with simple yet sensible

solutions for fostering economic growth and national productivity. It was the articulation of just such a program that helped to bring electoral success to avowed free-market advocates such as Margaret Thatcher in Great Britain (1979) and Ronald Reagan in the United States (1980).

RISING GOVERNMENT DEBT AND DECLINING ECONOMIC PERFORMANCE BEGAN CALLING EXISTING APPROACHES INTO QUESTION

The ideological shift was evident in Canada as well. In the interval between the Macdonald commission's appointment to its final report, the Progressive Conservatives displaced the Liberals. Under its businessman leader, Brian Mulroney, the Conservatives had been very adept at capitalizing on Liberal missteps on the campaign trail, the lacklustre national economic performance of the early 1980s, and the electorate's disenchantment with continual federal-provincial wrangling over taxation, energy policy, and constitutional matters in the past several years. Whereas the Liberal government's 1981 discussion paper on economic policy, *Economic Development Strategy for Canada in the 1980s*, had declared that the "Government of Canada must assume a leadership role in economic development," the Conservatives said that too much government was, in fact, part of the problem.

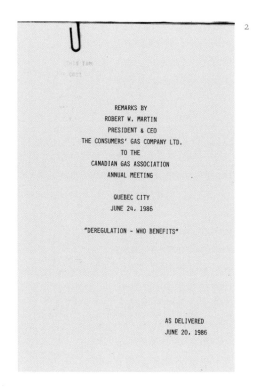

REMARKS BY
ROBERT W. MARTIN
PRESIDENT & CEO
THE CONSUMERS' GAS COMPANY LTD.
TO THE
CANADIAN GAS ASSOCIATION
ANNUAL MEETING

QUEBEC CITY
JUNE 24, 1986

"DEREGULATION - WHO BENEFITS"

AS DELIVERED
JUNE 20, 1986

In its 1984 discussion paper, *A New Direction for Canada: An Agenda for Economic Renewal*, the Conservative government asserted, "Government has become too big. It intrudes too much into the marketplace and inhibits or distorts the entrepreneurial process. Some industries are over regulated, others are over protected." During the next several years, it would bring forth a number of policies aimed at reducing the role of the state and increasing that of the market, including legislation to dismantle the NEP and the Foreign Investment Review Agency, and to implement the wide-ranging Canada-United States Free Trade Agreement.[3]

The policy changes that followed from the trend toward deregulation and freer trade would open up significant opportunities for the gas industry, creating additional possibilities for expansion at home and abroad. These changes would also create tremendous pressures for large-scale adjustments within what had long been a heavily regulated industry, during a period in which this industry was already struggling to come to terms with higher environmental and social costs. The result was that gas companies would have to become leaner and more responsive to the market and the community, while the CGA would need to radically reshape its entire organization. Though painful at times, these adjustments placed the industry in the strong position it is in today.

Toward deregulation in the energy sector: The Western Accord and the Halloween agreement

The modern era of energy services began with the signing of the Western Accord on 28 March 1985. Concluded by the federal government and the petroleum-producing provinces of Alberta, British Columbia, and Saskatchewan, the accord was intended to replace the heavily regulated NEP with a more market-oriented system of oil pricing and resource development. The two-tiered system of separate domestic and international pricing was abolished – leaving oil prices to be determined solely by the market – and federal taxes, export controls, and incentive programs were rolled back. These moves satisfied a major Conservative campaign pledge to reduce federal intervention in the provincial petroleum sector, a source of deep frustration for the West since the implementation of the NEP. Ostensibly, however, they were also undertaken in the belief that the market would provide a more equitable means of establishing prices and a more efficient method of allocating resources.[4]

On 31 October 1985, the accord was followed by a similar deregulation of the gas industry under the Halloween agreement between western provinces and the federal government. For years western provinces had complained that export restrictions and the monopoly position of the TransCanada pipeline were stifling access to markets. Under the existing system, gas exports could only proceed if there were at least thirty years of gas reserves for the domestic market. Furthermore, the gas that flowed to Eastern Canada through the TransCanada pipeline typically did so under large-volume, long-term contracts between major producers and the pipeline at one end and local distribution companies (LDCs)

3

2: A less regulated energy sector meant new challenges for the gas industry and the CGA. Consumers' Gas president Robert W. Martin, a former CGA chairman, addressed the complexities of deregulation in a paper delivered at the 1986 annual meeting in Quebec City. *CGA archive*

3: Despite a proliferation of third-party marketers, which fit with the goal of offering consumers more choice, infrastructure and delivery remained largely unchanged by deregulation. This meant gas from a storage facility such as Union's 150-billion-cubic-foot Dawn Hub, pictured, could be bought and sold by any number of third-party retailers. *Union Gas archive*

and the pipeline at the other. The relationship between producer and consumer was thus doubly mediated: firstly by TransCanada, which dominated the national transmission system; and secondly by local franchises, owned and operated by LDCs. This made it very difficult for smaller, independent producers in the West to break into the gas market and connect with smaller, independent consumers in the East. The agreement addressed these grievances by pledging the federal government to move toward relaxing export controls and "unbundling" the commodity portion of gas service from its delivery, thereby removing the physical and regulatory obstacles that stood in the way of allowing producers and consumers to link up more directly.[5]

To facilitate these linkages, the NEB ordered TransCanada to make pipeline capacity available for independent buyers and sellers. At first, such arrangements were hampered by the cost of financial penalties levied by TransCanada. These penalties, known as "unabsorbed demand charges," were applied in cases where it was necessary to make up for any shortfalls between contracts of independent buyers and sellers and the amount and cost of the gas that TransCanada would have normally delivered to LDCs under previously negotiated long-term contracts. Gradually, however, a series of regulatory decisions by the NEB and provincial regulators, along with the evolution of mechanisms for reselling any excess gas, cleared the way for the emergence of a more competitive system of gas sales during the period from the mid-1980s to the early 1990s.[6]

The elimination of export controls was the next step. After extensive study, the NEB decided such controls no longer served the public interest. After these controls were removed, in the late 1980s, Canada's gas reserves were reduced by the early 2000s to ten years' worth, one-third of the previous requirement and on a par with the US . From an economic standpoint, gas companies had been carrying much larger inventories than necessary throughout the postwar period, resulting in higher costs for investors and consumers alike. But from a historical perspective, this was simply part of the "price of doing business" at a time when the economic, scientific, and political frameworks of the industry were in the process of being established.[7]

At the provincial level, there was an "unco-ordinated convergence" with the regulatory changes at the federal level. In the mid-1980s, Ontario led the way by becoming the first province to allow all consumers to purchase their gas directly from a non-utility supplier rather than through their LDC. Consumers were thus free to shop around for the best possible prices. Initially, this change was primarily of benefit to the largest users of gas, for whom a small savings on the per-unit cost would mean a large savings on the total volume of gas purchased.[8]

IN THE MID-1980S, CONSUMERS WERE FREE TO SHOP AROUND FOR THE BEST POSSIBLE GAS PRICES AND PURCHASE DIRECTLY FROM A NON-UTILITY SUPPLIER

The opening of the gas market prefigured the appearance of "agents, brokers, and marketers" that purchased gas from producers on behalf of individual consumers, or groups of consumers. By buying gas in bulk, third-party distributors could negotiate better prices than individual consumers, and save these consumers the time and effort they would otherwise have to spend negotiating non-utility contracts on their own. Because third-party suppliers operated with low overhead and were not bound by long-term contracts, their prices were lower than those offered by the LDCs. As an added service, these suppliers could also offer their customers long-term contracts with a set price. This gave consumers the option of price stability, while allowing third-party suppliers to earn additional profits if they could secure gas at a lower price than what had been negotiated with their customers. LDCs, by contrast, could only charge the going market price for gas, plus a regulated delivery charge. Their earnings were determined by a return-on-investment formula that took into account their capital investment in the physical infrastructure that actually delivered all of the gas distributed into their franchise areas – regardless of whether or not the gas had been ordered by independent consumers, third-party suppliers, or their own consumers.

Third-party suppliers tended to entrench deregulation in the gas market, a process driven by a combination of business imperatives, consumer interest, and ideological trends. As brokers sought to expand their businesses, they proceeded to sign up some of the smaller consumers in the commercial and residential sectors, such as hotels, retail outlets, restaurants, and apartment buildings. Eventually there was an interest in residential heating and cooking markets, too. For policymakers, third-party services paired nicely with the prevailing taste for expanding the role of market mechanisms in an effort to increase efficiency and consumer choice. As a result, in the late 1990s and the early 2000s, provinces such as Ontario, Quebec, and Manitoba put forth legislation to make it easier for brokers to expand their presence.[9]

Like most such changes, the move toward deregulation has had both costs and benefits. In the period from 1985 to 1991, as a result of the glut of supply released by deregulation, the average wellhead price of natural gas in Canada fell from approximately $2.20 to $1.20 per gigajoule. Then, as demand and supply rebalanced throughout the remainder of the 1990s, prices increased to between $3 and $4 per gigajoule. Since the early 2000s, the supply overhang has been absorbed and there have been three sharp spikes in gas prices – in the winters of 2000, 2003, and 2005 – and prices have increased to about $8 per gigajoule. With an open and balanced gas market, price fluctuations are tied to economic fundamentals on the demand side, not the vagaries of supply that were often reflected in the gas bills of industrial, commercial, and residential consumers. Expanded markets and higher prices also promoted many additional investments. From the early 1990s to the early 2000s, the average number of gas wells drilled per year in Canada increased to more than 14,000 from less than 3,000. During the same period, the transmission network was significantly expanded as well, with transmission companies investing more than $8 billion between 1995 and 2000 alone.[10]

When compared to the experience of other industries that underwent deregulation, such as electricity and airlines, these results are a macroeconomic success. Prices have risen because of higher demand. But they have done so without causing major economic disruption, thanks in part to the extensive supplies of gas that had been built up under the thirty-year reserve requirement that was in place prior to deregulation. Moreover, new investments

in exploration and transmission since the early 1990s have improved the gas system's responsiveness to changes in demand and supply, thus creating conditions for more price stability in the future.[11]

For LDCs, deregulation meant big changes to their business model. No longer could local gas utilities continue to provide the full range of "final end" services to their customers, such as the sale of gas appliances. Under deregulation, policymakers felt that the role of these companies should be restricted to that of being the distributors of the commodity. Otherwise, they feared utilities might be tempted to "cross-subsidize" the non-utility portion of their business, along with the additional support services required in the form of computer systems and sales outlets, with assets paid for by the regulated rates charged to customers. To prevent this from happening, regulators required LDCs to sell their non-regulated businesses.[12]

At the residential level, many consumers have found the choices offered by deregulation to be annoying and confusing rather than liberating and helpful. In highly regulated markets, such as Saskatchewan, business-customer relations are much simpler. All consumers buy gas at a fixed price from SaskEnergy, the provincially owned distributor; all customers have a limited range of options and rates. In deregulated jurisdictions, such as Ontario, Alberta, and British Columbia, consumers have more choice. They can purchase gas either from an LDC or a third party. If they buy their gas from an LDC, they are charged the commodity rate, plus a delivery charge based on a return-on-investment formula; if they buy their gas from a third party, they are charged the commodity rate (based on the market price), plus the established delivery charges; or an agreed-upon rate for the commodity (based on a short-, or long-term contract), plus delivery. In each case, consumers balance their tolerance for risk, their need for price stability, and their desire for an ongoing relationship with a supplier.[13]

4: More than $8 billion was invested in the transmission network between 1995 and 2000. Such improvements have created a vastly expanded coverage area – one that far exceeds the system of just a few decades earlier. Map circa 1964. *Canadian Gas Journal*

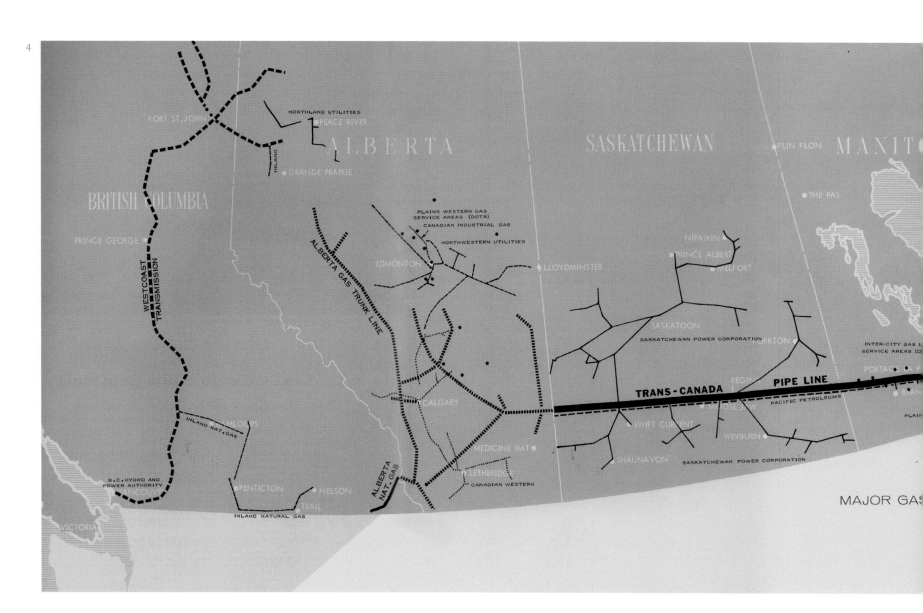

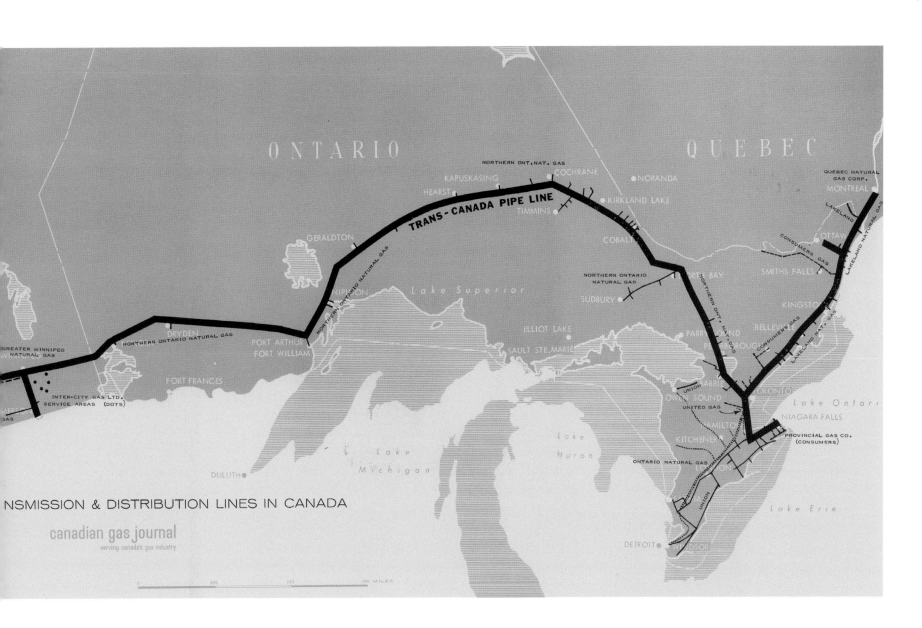

ONTARIO

QUEBEC

NORTHERN ONT. NAT. GAS

KAPUSKASING

HEARST

COCHRANE

NORANDA

QUEBEC NATURAL
GAS CORP.

MONTREAL

TRANS-CANADA PIPE LINE

KIRKLAND LAKE

LAKELAND

TIMMINS

GERALDTON

COBALT

CONSUMERS GAS

OTTAWA

NORTHERN ONTARIO NATURAL GAS

Lake Superior

NORTHERN ONTARIO
NATURAL GAS

NORTH BAY

NORTHERN ONT. NAT. GAS

SMITHS FALLS

NIPIGON

SUDBURY

KINGSTON

DRYDEN

ELLIOT LAKE

PARRY SOUND

BELLEVILLE

NORTHERN ONTARIO NATURAL GAS

PETERBOROUGH

PORT ARTHUR
FORT WILLIAM

SAULT STE. MARIE

CONSUMERS GAS

LAKELAND NAT. GAS

GREATER WINNIPEG
NATURAL GAS

UNION

BARRIE

TORONTO

FORT FRANCES

OWEN SOUND

Lake Ontario

INTER-CITY GAS LTD.
SERVICE AREAS (DOTS)

UNITED GAS

NIAGARA FALLS

HAMILTON

Lake
Huron

PROVINCIAL GAS CO.
(CONSUMERS)

KITCHENER

EMERSON
GAS

Lake
Michigan

ONTARIO NATURAL GAS

LONDON

DULUTH

UNION

Lake Erie

NSMISSION & DISTRIBUTION LINES IN CANADA

DETROIT

WINDSOR

canadian gas journal
serving canada's gas industry

0   100   200   300 MILES

There have been other issues as well. Consumer frustration in deregulated provinces has been heightened by the market-driven price volatility that has, at times, driven prices up and down in many jurisdictions. It has also been aggravated by the questionable sales practices of a few third-party suppliers, commission-based agents that misrepresented which companies they were affiliated with or signed up customers without their consent. And although the situation has been improved by new regulations and a better understanding of the market on the part of consumers, deregulation remains, in some ways, a tarnished idea. At least some gas utilities have also wondered whether it might be advantageous to return to the business of selling gas appliances and equipment, if only for the purpose of actively promoting their product. In the long run, though, the adoption of a more market-oriented system of gas services should assist in promoting the efficient pricing of natural gas, expanding its use, supporting the development of new sources of supply, and holding down energy costs. As well, market-based pricing of gas forces consumers to consider the real costs of their energy choices, a factor directly linked to the other primary driver of energy policy from the late 1980s onward, the environment.[14]

STRAIN ON THE ENVIRONMENT WILL ONLY INCREASE AS GLOBAL ENERGY DEMAND CLIMBS TO NEW HEIGHTS

The Brundtland report:
The challenge of sustainability

Throughout the 1980s and 1990s, the world's demand for energy showed few signs of abating. From 1980 to 2003, the world's energy consumption increased to 421 quadrillion British Thermal Units (BTUs) from 283 quadrillion BTUs. This expansion occurred across residential, commercial, industrial, and transportation markets. Up until very recently, much of it has been concentrated in the most developed nations of the Northern Hemisphere. More and more, however, the significant increases have been concentrated in the developing nations of Asia, Africa, and South America. Growth in these regions flows from the diffusion of wealth and technology. These developments have been enormously beneficial to large numbers of people, but they have also put greater and greater strain on the natural environment, strain that will only increase as global demand moves toward a projected 665 quadrillion BTUs by 2025.[15]

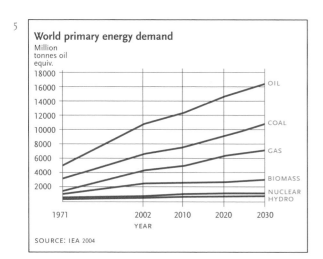

5

World primary energy demand
Million tonnes oil equiv.

SOURCE: IEA 2004

The United Nations held the first Conference on the Human Environment in Stockholm, Sweden, in 1972, which brought together representatives of 113 countries to "consider the need for a common outlook and for common principles to inspire and guide the peoples of the world in the preservation and enhancement of the human environment." Later that year, the United Nations Environment Program was formed to "act as the world's environmental conscience." Both were indicative of growing concerns about the environmental costs of human development.[16]

In 1983, these concerns led to the establishment of the World Commission on Environment and Development, mandated to investigate the environmental challenges of the next twenty years and to consider ways by which the international community could meet these challenges. Gro Harlem Brundtland, a medical doctor and former prime minister of Norway, chaired the commission. Its report, released in 1987, developed the broad principle of "sustainable development," which it defined as "development that meets the needs of the present without compromising the ability of future generations to meet their own needs."

The Brundtland report recognized the indivisibility of economic development, social equity, and the environment. It pointed out that just as economic advancement has environmental implications, the degradation of the environment has economic costs. It therefore recommended that the benefits of future economic growth should be shared more equitably with the nations of the developing world, so that the peoples of these countries would not need to inflict undue harm upon their environments in order to survive. To underwrite this process, it further called for the development of new resources and the improved management of existing ones in all regions of the world. The report was followed by the call for the Rio Earth Summit of 1992, which was followed by many conferences, agreements, and progress on environmental issues.[17]

Canadians shared in the international anxiety about the state of the environment. During the last two decades of the twentieth century, political parties of almost all persuasions made environmental issues a larger and larger part of their policy agendas. Inside the federal bureaucracy, the position of Environment Canada has been continually enhanced as a result. Between 1999 and 2005 alone, for instance, its budget grew to a peak of $958.7 million from $559.4 million, as did its involvement in the diverse policy areas that fall within its purview, including energy, industry, trade, and international relations.[18]

6: The Brundtland report, published in 1987, was the first document to flesh out the term 'sustainable development,' which puts equal emphasis on environmental, social, and economic health. The report stresses that each aspect is part of a continuum – a continuum, moreover, that is vital to the planet's survival. *United Nations*

6

UNITED NATIONS

General Assembly

A

Distr.
GENERAL

A/42/427
4 August 1987
ENGLISH
ORIGINAL: ARABIC/CHINESE/ENGLISH/
FRENCH/RUSSIAN/SPANISH

Forty-second session
Item 83 (e) of the provisional agenda*

DEVELOPMENT AND INTERNATIONAL ECONOMIC CO-OPERATION: ENVIRONMENT

Report of the World Commission on Environment
and Development

Note by the Secretary-General

1.   The General Assembly, in its resolution 38/161 of 19 December 1983, inter alia, welcomed the establishment of a special commission that should make available a report on environment and the global problématique to the year 2000 and beyond, including proposed strategies for sustainable development. The commission later adopted the name World Commission on Environment and Development. In the same resolution, the Assembly decided that, on matters within the mandate and purview of the United Nations Environment Programme, the report of the special commission should in the first instance be considered by the Governing Council of the Programme, for transmission to the Assembly together with its comments, and for use as basic material in the preparation, for adoption by the Assembly, of the Environmental Perspective to the Year 2000 and Beyond.

2.   At its fourteenth session, held at Nairobi from 8 to 19 June 1987, the Governing Council of the United Nations Environment Programme adopted decision 14/14 of 16 June 1987, entitled "Report of the World Commission on Environment and Development" and, inter alia, decided to transmit the Commission's report to the General Assembly together with a draft resolution annexed to the decision for consideration and adoption by the Assembly.

3.   The report of the World Commission on Environment and Development, entitled "Our Common Future", is hereby transmitted to the General Assembly. Decision 14/14 of the Governing Council, the proposed draft resolution and the comments of the Governing Council on the report of the Commission can be found in the report of the Governing Council on the work of its fourteenth session. 1/

----
   *   A/42/150.

87-18467   2499h (E)   /...

7: Environmental issues extend beyond the tidy boundaries that divide nation-states, increasing the need for international co-operation. For their part, gas industries in Canada, Mexico, and the US have collaborated to find solutions. The agenda pictured is from a 1994 meeting at Mexico City. *CGA archive*

8: This CGA pamphlet, circa 2000, highlights some of the efforts undertaken to make natural gas part of a responsible energy portfolio in an era of growing concern over rising greenhouse-gas levels and climate change. *CGA archive*

9: Meeting the goal of sustainability requires a balance between making use of existing infrastructure, in which billions have been invested, and looking to more efficient, greener technology. Terasen's Triple Point metering system is pictured. *Terasen Gas*

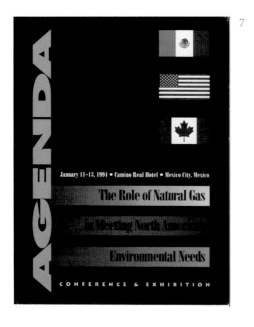

7

8

CANADA STILL FACES A SIGNIFICANT CHALLENGE IN REDUCING ITS GREENHOUSE-GAS EMISSIONS

In some respects, the interests of the gas industry have long paralleled those of the environmental movement. Beginning in the early twentieth century, the pressures of competition from the electrical industry prompted gas appliance manufacturers to constantly improve the efficiency of their products. The more cost-effective gas appliances could become, the more likely that consumers would be persuaded to purchase and employ them. Indeed, the efficiency of cooking and home heating with gas has been, and remains, among its main selling features and benefits.

Nonetheless, Canada still faces a significant challenge in reducing its greenhouse-gas emissions. On a per-capita basis, it is among the highest emitters in the world. This reflects both its resource endowments and its patterns of energy consumption. Canada possesses a wealth of natural resources in the form of coal, oil, and natural gas, all of which produce greenhouse-gas emissions as they are converted into energy. As a result of its economic wealth, it also has many citizens who possess and enjoy the use of all manner of energy-consuming vehicles, climate-control appliances, telecommunications systems, and other electronic devices and gadgets.

Since the challenges are systemic, the solutions must be systemic as well. Canada has invested billions in its existing energy systems. This infrastructure cannot be reshaped or discarded in an instant without enormous economic and social costs. At the same time, piecemeal changes targeted at small portions of the system will produce only piecemeal results. What is needed is a long-term, comprehensive strategy.

The gas industry sees itself as part of this strategy. Natural gas is a "clean-burning" fuel that emits much less greenhouse gas than either coal or oil, and thus affords opportunities for immediately reducing greenhouse-gas levels through fuel-switching wherever practicable. The CGA also continues to work with the federal government and other stakeholders to promote better energy choices on the part of consumers, while recognizing that the means for doing so must take into account the fact that consumer behaviour is shaped not only by environmental factors but also those relating to efficiency, safety, and reliability. A strategy that incorporates these elements offers great potential for progress on the environmental front, though such progress will take strong and sustained collaborative efforts on the part of businesses, governments, and consumers.[19]

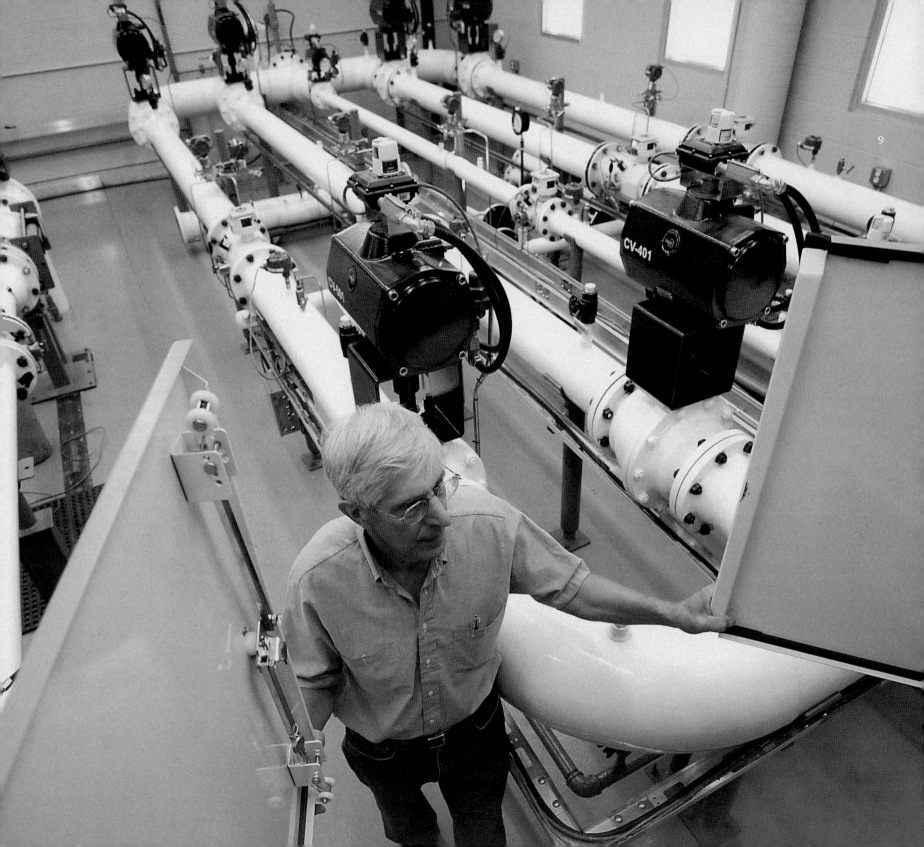

10, 11: By the 1980s, testing and certifying appliances had become a complex affair. Regulations were added each year, while existing standards changed to reflect new technology. The CGA finally sold the testing and certification division in the late 1990s to the Canadian Standards Association, ending the era of the CGA's seal of approval. *CGA archive*

12: After the CGA stopped dealing with the retail marketplace and end-use products, the Canadian Gas Research Institute, with its emphasis on improving the efficiency of gas appliances, was an expense that was hard to justify. Information booklet circa 1977. *CGA archive*

10
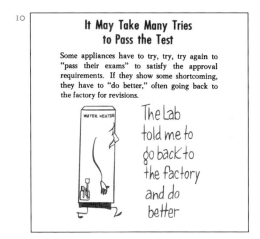

It May Take Many Tries to Pass the Test

Some appliances have to try, try, try again to "pass their exams" to satisfy the approval requirements. If they show some shortcoming, they have to "do better," often going back to the factory for revisions.

The Lab told me to go back to the factory and do better

11

THE STORY OF THE CANADIAN GAS ASSOCIATION'S APPROVAL SEAL

The CGA:
From trade association to think-tank

By the end of 1980s, the CGA had become a large, national business association with a staff of more than 140 employees and a hand in almost every aspect of industry-related research, statistics, public relations, appliance testing, and business-government relations. Most of these functions had been taken on in the four decades since the end of the Second World War, when the association was expanding to take on the challenges of building up the industry and thus providing a range of services that would not otherwise have been available. By the late 1990s, however, it was no longer quite as obvious that all of these functions should, or could, be best provided through the association itself. Some functions were duplicated elsewhere; others could be better served by different organizations. All had become more expensive to provide. As the CGA board saw it, the time had come, once again, to rethink the nature of the association.

The decision concerning what should be done about the CGA's appliance testing and certification division was one of the biggest and toughest that the board faced. Beginning in the early 1960s, this division had made significant contributions to maintaining high standards of equipment safety and building up the market for gas services. It also employed more than two-thirds of the association's workforce. Yet the costs of the CGA laboratories had escalated considerably in recent years. Every year, the number of regulations and the types of equipment to be tested and certified multiplied, and this meant higher costs for everything from labour to equipment. Moreover, the development of equipment standards was fast becoming an international concern, as trade barriers and transportation costs declined and manufacturers sought out bigger and broader markets to remain competitive. Even as the CGA continued to produce its own "national standards," someone would still have to ensure that these were congruent with those being produced elsewhere in order to enable Canadian manufacturers and suppliers to remain competitive on the international stage.[20]

For all of these reasons, the testing and certification business was sold off in the late 1990s. It was purchased by none other than the Canadian Standards Association, the same organization the CGA had thought too expensive and too unresponsive to its needs to serve as one of its testing facilities back in the 1950s. With international connections and standards as the CSA's main business, however, it was clear that it was now better positioned to become the national centre for appliance testing and certification in the changed conditions of the late 1990s. Some of the staff from the CGA's testing division joined the CSA, some took early retirement, and some moved on to other opportunities.

The late 1990s also witnessed the closing of the CGA's Canadian Gas Research Institute. When it was initiated in the mid-1970s, the CGRI addressed an important aspect of gas research that was not being well served at the time: that of finding additional applications for gas services and making gas equipment more efficient. As in the case of appliance testing, however, the CGA had to evaluate its role in research on end-use products to reflect the new market realities. Retail-market deregulation diminished the role of gas utilities' contact with customers' appliances and introduced new players. It made sense to allow these new players to assume the costs, risks, and responsibilities associated with product development.[21]

It was much the same with other aspects of the CGA's activities. From the early 1910s to the early 1980s, the CGA had made extensive efforts to collect and disseminate statistical information on the gas industry to allow its members to make well-informed business decisions. Particularly in the early years, CGA statistics were among the most reliable and accessible. As the gas industry gained prominence in the national economy, other organizations took on a larger role. By the late 1980s, industry statistics were being gathered by all manner of public and private organizations, and this reduced the need for the CGA to provide this service.

12

Industrial research on a shoestring
The CGA is doing quite nicely thank you very much

TUCKED AWAY on a quiet street in one of those manicured industrial parks in suburban Toronto is a small and little known research group that is saving Canadians untold hours of inconvenience and the industry it serves literally millions of dollars. And the astonishing fact is, they're doing it on what many researchers would consider an impossible budget.

The group, consisting of some eight individuals, sort of backed into existence about five years ago as the child of the Approvals Division of the Canadian Gas Association, a corner of whose laboratory in Don Mills it now occupies.

In the technical area, the CGA has been mainly known for the work of the Approvals Division in the testing, evaluation and certification of gas-fired apparatus for home and industry. But in recent years, the association has been increasing its activity in contract research with the result that it is developing a competence that is becoming recognized throughout North America and Europe.

In contrast to many of Canada's industrial-oriented research laboratories today, the CGA laboratory is electric (that may be the wrong word) with activity. Morale is high, and on going through the lab, the researchers are constantly pointing with no little pride to "hoked-up" pieces of equipment and explaining how they built them themselves at next to no cost and how this equipment enabled them to carry on with a certain research project. And it is just this attitude and resourcefulness that has enabled these researchers to bring to fruition research contracts that vastly better equipped labs have been unable to fulfill on far larger budgets.

The Canadian Gas Association was formed in 1907 by 12 men from a handful of companies when the industry was in its infancy. Today it speaks for a $6 billion industry in Canada with more than 650 corporate and individual members drawn from the utilities and transmission companies, producers, pipeline contractors and manufacturers of gas-fired appliances and equipment.

Actually, as Bill Dalton, CGA's managing director, points out, the association's research efforts grew out of the realization that service costs were rising at alarming rates. Until a few years ago most of these costs had been borne by the industry, but lately the rapidly rising costs of service calls proved too heavy a financial burden for the industry to absorb alone.

One such problem was the cracking of heat exchangers in forced air furnaces. Although the danger of flue gases escaping into the building proved to be minimal, the cracks in some cases altered the combustion conditions in the furnace. Research on this problem has now been completed with a resultant saving to the industry of millions of dollars. Another problem the industry was having involved corrosion in hot water tanks. This research project is now being wrapped up and also promises to pay high dividends. Still another project deals with failing thermocouples, used to monitor pilot flames. When the thermocouple fails, the furnace shuts off and requires a service call to restart it.

Since such failures are not uncommon and the service calls cost a minimum of $7 each, the problem represents a considerable cost to the utilities who have become in recent years heavily involved in the servicing of their customers' equipment. While this research program is still under way, it has already produced information for the redesign of one manufacturer's thermocouple as well as a small device the service man can use to test the thermocouple in the field when he is called out on some other service problem.

While the association's original research efforts could at best be described as modest, with every successfully completed project it's finding it less difficult to obtain industry backing. The research group has repeatedly demonstrated that with what seems to be very little expenditure it can save association members a great deal of money.

But the story of R&D activities at the CGA is not just how much money it has saved the industry or how much inconvenience it has saved the public, it is also a story of how the work was done on a

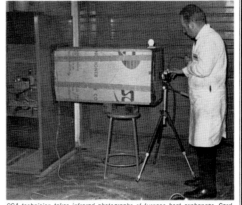

CGA technician takes infrared photographs of furnace heat exchanger. Cardboard box minimizes effects of stray radiation and allows photographs to be taken during the day in a lighted laboratory.

THE LATE 1990S ALSO WITNESSED THE CLOSING OF THE CGA'S CANADIAN GAS RESEARCH INSTITUTE

13: Print materials produced by the CGA in the 1990s reflect the association's diverse and adaptive functions in the gas industry. *CGA archive*

14: Having adopted a policy-oriented mandate, the CGA sought a location more suitable to its kind of work than the Don Mills offices, which had been situated amid a manufacturing and utilities hub. Company letterhead, 1988. *CGA archive*

13

Between the early 1960s and the early 1990s, the CGA had been deeply involved in the promotion of gas services through the production of newspaper advertising, consumer information pamphlets, television commercials, stuffed toys, and films. Toward the end of the 1990s, though, the dramatic expansion of the demand for gas seemed to make some of this work unnecessary. The removal of the utilities from the "white goods" end of the market as a result of deregulation had a similar effect, and so too did a shift among Canadian manufacturers and distributors into other types of gas equipment aimed more at the utilities themselves. Perhaps there was also some sense that gas was by then a widely accepted source of energy, its benefits well known to consumers, contractors, and governments. This shift called for a streamlined public-relations division.

14

TESTING LABORATORIES

CANADIAN GAS ASSOCIATION

55 SCARSDALE ROAD

DON MILLS • ONTARIO.

CERTIFICATION SERVICES. RESEARCH. SPONSOR OF CSA STANDARDS. GOVERNMENT LIAISON.

Some searching questions still lingered at the turn of the twenty-first century. Was there a need for the CGA? If so, what was its function? Could the same ends be achieved through another organization?

After much consultation and reflection, it was readily apparent that a reorganized association was vital to the industry. The community would benefit from the association's assistance in managing the constantly changing economic, environmental, and social dimensions of the industry's regulatory framework, and the membership would benefit from the association's ability to communicate with other stakeholders on behalf of the industry as a whole. There was also no other organization that could serve these particular functions as well as the CGA had been doing for close to a century.

Since the early 1960s, the CGA had been headquartered in Don Mills, close to many appliance and equipment manufacturers, the laboratories and expertise of the Ontario Research Foundation, and the headquarters of several major utilities. But with the testing and certification division gone, and a policy and communications focus emerging, not to mention a large portion of the membership based in the West, it was felt that it was time to move, and Calgary and Ottawa were the leading options. In the end, Ottawa was selected, because it best fit with a policy-oriented mandate. The offices at 55 Scarsdale Road in Don Mills were sold shortly after the appliance-testing divisions were sold in the late 1990s. After a couple of relocations within the Toronto area, the CGA established its head offices at 350 Sparks Street, in the nation's capital.

As the association evolved, the CGA board and management reoriented the association toward collecting, disseminating, and analysing industry-related information; bringing members together to exchange ideas; and working with government and other stakeholders to locate the right energy-policy framework. After a century of fuelling industry progress, the CGA had returned to its roots.

Outlook

Previous: View of Terasen Gas's Southern
Crossing pipeline during construction
in November 2000. *Terasen Gas*

Industry, trade, and political associations are an integral part of Canadian history. By aggregating the opinions of economic, social, and political groups, they have been essential to the political process. By establishing common trade standards and practices, they have contributed to the development of industries and professions. And by bringing individuals together under common causes, they have created a greater sense of national identity. The Canadian Gas Association exemplifies these contributions. As the voice of suppliers, manufacturers, and other interested parties, the CGA has profoundly affected the gas industry, as well as national energy policy. Throughout the association's one-hundred-year history, adapting to changing conditions has been a crucial part of its success along the way.

When the CGA was established, the gas industry was fragmented. Gas manufacturers of various sizes served numerous communities throughout Canada, while natural gas companies served only a few urban centres and rural hamlets in Ontario and Western Canada. Many of these companies faced stiff competition – from relatively inexpensive coal, hydroelectricity, or both. Indeed, the loss of the gas industry's lighting business to electrical competitors was a catalyst for the formation of CGA by a small number of Ontario-based gas manufacturers in 1907. Through collective action, the CGA's founders hoped to find new ways to expand the industry.

In this early phase of development, the association formed crucial linkages among the constituent elements of the gas business; its official journal kept CGA members up to date on the latest developments in technology, management practices, and customer service; and its annual conferences did much the same, providing opportunities for forging the personal connections that are so important to successful collaboration in any human endeavour. Not least, the organization also provided an efficient means for uniting and expressing the disparate interests of the industry as a whole.

After the Second World War, the association fostered further expansion of the gas industry by filling numerous gaps in research, regulation, and communications. From the 1950s to the 1970s, entire divisions of the CGA were devoted to gas statistics, public relations, and human resources. The founding of the association's own appliance testing and certification laboratories was among the most significant of these efforts. From the early 1960s to the early 1980s, these facilities greatly eased the process of generating national standards for gas equipment and appliances, as the industry became national rather than regional. With the evolution of continental, and then global, gas markets during the late twentieth and early twenty-first century, the association again proved its worth by assisting members, government regulators, and consumers to make the transition.

What role can this century-old Canadian association play in the new global environment? A vital but different one. For much of the twentieth century, the association was preoccupied with nurturing the industry itself. This was appropriate at a time when the full potential of natural gas was only dimly perceived, the political and physical infrastructures of the industry were poorly developed, and the competitive challenges were multiple, varied, and persistent. But with natural gas supplying roughly 31 percent of Canada's energy needs and widely recognized as an efficient and cleaner-burning fuel, the need for the CGA's economic, intellectual, and political spadework seems far less obvious.

Yet many challenges remain. We live in a world in which human development is having a massive impact on the natural environment. This raises questions about the definition of human progress, and how such progress can be sustained without, as the Bruntland report put it, "compromising the ability of future generations to meet their own needs." This, in turn, raises a host of questions about the overall security of energy supplies, as well as the optimal mix of alternative forms of energy, now and in the future.

Natural gas will be part of this mix. In Canada alone, Natural Resources Canada estimates there are about 594 trillion cubic feet of proven and unproven gas resources available. This is the equivalent of 100 years of supply at current national consumption rates. Even as demand increases, these figures could be extended, so long as technology becomes more efficient, consumers adopt more measures to conserve energy and reduce costs, and additional sources of supply are developed. Beyond that, there are also untapped forms of gas, such as methane hydrates, that will further extend supplies.[1]

In the coming decades, other fossil fuels will maintain their presence, as will conventional alternatives, such as nuclear power. There will also be a growing reliance on renewable sources of power, such as wind, and on emerging technologies such as hydrogen fuel cells. The central challenges will be to find and achieve the best possible balance between economic efficiency and environmental responsibility. Doing so will require gathering a great deal of information and analysis, generating a strong political will for change, and promoting the right consumer choices. It is in these areas that the Canadian Gas Association will make its mark in the next one hundred years.

Notes

Chapter One

1 "Local Benefits of Policy," Toronto *Star*, 21 November 1907, 8; and "There Cannot Be Too Much Light," Toronto *Globe*, 21 November 1907, 1.

2 F.H. Leacy, M.C. Urquhart, and K.A.H. Buckley (eds.), *Historical Statistics of Canada*, 2nd ed. (Ottawa: Statistics Canada, 1983), A1, A350, and M301-309; *Canada Yearbook* (Ottawa, 1908), 132, 149, 383, and 438; and R.C. Brown and Ramsay Cook, *Canada 1896-1921: A Nation Transformed* (Toronto: McClelland and Stewart, 1974), 1-6 and 49.

3 On the early uses of coal, oil, and natural gas, see Peter McKenzie-Brown, Gordon Jaremko, and David Finch, *The Great Oil Age: The Petroleum Industry in Canada* (Calgary: Detselig Enterprises, 1993), 24-5; Lawrence F. Jones, "Oil and Civilization," in James D. Hilborn, *Dusters and Gushers: The Canadian Oil and Gas Industry* (Toronto: Pitt Publishing, 1968), 2-4; and William Kilbourn, *Pipeline: Trans-Canada and the Great Debate, a History of Business and Politics* (Toronto: Clarke, Irwin and Co., 1970), 3-4 and 6-8.

4 D.G. Creighton, *The Commercial Empire of the St. Lawrence* (Toronto: Ryerson, 1937); and *Census of Canada* (Quebec, 1853), Appendix C.

5 On Albert Furniss, see Christopher Armstrong and H.V. Nelles, *Monopoly's Moment: The Organization and Regulation of Canadian Utilities, 1830-1930* (Philadelphia: Temple University Press, 1986), 13-15; and Fernand Ouellet, "Masson, Joseph," *Dictionary of Canadian Biography* (hereinafter, *DCB*) (Toronto and Montreal: University of Toronto/Université Laval, 2000).

6 Creighton, *Empire of the St. Lawrence*, 358-63; and Kenneth Norrie and Douglas Owram, *A History of the Canadian Economy*, 2nd ed. (Toronto: Harcourt Brace Jovanovich, 1991), 207-15.

7 Armstrong and Nelles, *Monopoly's Moment*, 14 and 21-2; and Edward J. Tucker, *The Consumers' Gas Company of Toronto* (Toronto: Consumers' Gas Company, 1923), 57.

8 Tucker, *The Consumers' Gas Company*, 53; Armstrong and Nelles, *Monopoly's Moment*, 22-5; and "The 'White Way' in Canada," Science and Technology Museum of Canada, www.sciencetech.technomuses.ca/english/collection/electr4.cfm, 6 January 2006.

9 See Christopher Vodden, "No Stone Unturned: The First 150 Years of the Geological Survey of Canada," www.gsc.nrcan.gc.ca/hist/150_e.php, 22 January 2006.

10 Kilbourn, *Pipeline*, 8; and McKenzie-Brown, Jaremko, and Finch, *The Great Oil Age*, 32.

11 *Who's Who in Canada* (Toronto: International Press, 1923); and Victor Lauriston, *The Blue Flame of Service: A History of the Union Gas Company and the Natural Gas Industry in Southwestern Ontario* (Chatham: Union Gas Company, 1961), 10-12.

12 On the Essex Gas field, see Lauriston, *The Blue Flame of Service*, 14-15.

13 G.W. Allen to K.R. Boyes, "The Canadian Gas Association: Its Rise and Development," 7 April 1932, CGA archive, Box D.

14 CGA, Proceedings (1908), 14 and 19-22.

15 On the functions and strategies of political associations, see Stephen Brooks and Andrew Stritch, *Business and Government in Canada* (Scarborough: Prentice-Hall, 1991), 221-2; and W.T. Stanbury, *Business and Government Relations in Canada: Influencing Public Policy* (Scarborough: Nelson, 1993), 120-2 and 125-7. Also, see "Convention of Gas Men," *Globe and Mail*, 27 June 1908; and "The Gas Men's Meeting," *Montreal Gazette*, 30 June 1909.

16 CGA, Proceedings (1908), 7 and 20-2.

Chapter Two

1 "Eugene Coste," *The Canadian Mining and Metallurgical Bulletin*, (Aug/1840); McKenzie-Brown, Jaremko, and Finch, *The Great Oil Age*, 32-3; Len Stahl, *A Record of Service: The History of Western Canada's Pioneer Gas and Electric Utilities* (Edmonton: Canadian Utilities Limited, 1987), 2.

2 See Eugene Coste, "Natural Gas in Ontario," *Canadian Mining Institute, Proceedings* (1900); "The Volcanic Origin of Oil," *Canadian Mining Institute, Proceedings* (1903); "Natural Gas and Oil Possibilities in Ontario," *Natural Gas and Petroleum Association of Canada (NGPA), Proceedings* (1923); McKenzie-Brown, Jaremko, and Finch, *The Great Oil Age*, 21-2; and Lauriston, *Blue Flame of Service*, 13.

3 On the discovery and development of the Bow Island field, see Stahl, *A Record of Service*, 2-4; and Ed Gould, *Oil: The History of Canada's Oil and Gas Industry* (Victoria: Hancock House, 1976), 59-63.

4 "Directory of Gas Companies," *Intercolonial Gas Journal*, Dec 1914, 460-1.

5 D.A. Coste, "History of Natural Gas in Ontario," *Intercolonial Gas Journal*, Nov 1919, 423-30.

6 On the Acme Oil Company and the development of the Tilbury oil field, see Lauriston, *Blue Flame of Service*, 21-3; Coste, "History of Natural Gas in Ontario," 426; and Win Miller, *Union Gas: The First Seventy-Five Years* (Chatham: Union Gas, 1986), 2.

7 Lauriston, *Blue Flame of Service*, 22-3; and the Judicial Committee of the Privy Council, Canadian Reports. Appeal Cases (Toronto: Arthur Poole and Company, 1912).

8 Lauriston, *Blue Flame of Service*, 14, 23-30, and 36; and Coste, "History of Natural Gas in Ontario," 426.

9 Coste, "History of Natural Gas in Ontario," 423-7.

10 Miller, *Union Gas*, 2-3.

11 On the formation of Union Gas, see Miller, *Union Gas*, 2-3; and Lauriston, *Blue Flame of Service*, 36-8.

12 Regarding the Southern and Union companies in the 1910s, see Lauriston, *Blue Flame of Service*, 36-40.

13 J.R. Colton and R.R. Palmer, *A History of the Modern World*, 6th ed. (New York:

McGraw-Hill, 1984), 660-71. Also see, Paul Kennedy, *The Rise and Fall of the Great Powers: Economic Change and Military Conflict from 1500 to 2000* (London: Fontana Press, 1988), 249-354.

14 "Cooking with Gas on a Mammoth Scale," *Intercolonial Gas Journal*, Jan 1916, 10; and "Helping Kitchener to Finish the War," *Intercolonial Gas Journal*, Jan 1916, 11.

15 On Edmonton and Viking, see ATCO archives, Northwestern Utilities, *History and Information Book*; Stahl, *Record of Service*, 17-18; and Gould, *Oil*, 66-7.

16 Paul Reading, "Oil Pathfinders at Turner Valley Had 10 Difficult Years," *Calgary Herald*, 12 April 1929; Kilbourn, *Pipeline*, 10-11; and McKenzie-Brown, Jaremko, and Finch, *The Great Oil Age*, 36.

17 Lauriston, *Blue Flame of Service*, 48.

18 Lauriston, *Blue Flame of Service*, 52-3; and Coste, "History of Natural Gas in Ontario," 428.

19 Lauriston, *Blue Flame of Service*, 53-7; Ontario Municipal and Railway Board, *Annual Report* (Toronto: King's Printer, 1918); and Natural Gas Act, *Statutes of Ontario* (Toronto: King's Printer, 1918).

20 On the formation of the NGPA, see Lauriston, *Blue Flame of Service*, 59-61; Victor Lauriston, "Way Back When ... Gas Men Joined to Help Industry," *Gas Lines*, May 1951, 1-2; and "Report of the Proceedings of the Natural Gas and Petroleum Association of Canada," *Intercolonial Gas Journal*, July 1919, 247-54.

21 "Report of the First Annual Convention, Natural Gas and Petroleum Society of Canada," *Intercolonial Gas Journal*, Oct 1919, 367-79.

21 *Intercolonial Gas Journal*, Oct-Nov 1919.

22 "Plans for the Amalgamation of the American Gas Institute and the National Commercial Gas Association Rapidly Approaching Completion," *Intercolonial Gas Journal*, June 1918, 135; and "American Gas Association Organized," *Intercolonial Gas Journal*, July 1918, 254-5.

23 "Proposed Amendments to the Constitution Pertaining to the Membership in the Canadian Gas Association," *Intercolonial Gas Journal*, Jan 1919, 25-8; and "Address of President C.C. Folger Before the Twelfth Annual Convention, Canadian Gas Association, Niagara Falls, Ontario," *Intercolonial Gas Journal*, Sept 1919, 329-31.

24 DBS, *Canada Yearbook* (1908), 136-7; DBS, *Canada Yearbook* (1911), 75; DBS, *Canada Yearbook* (1920), 255 and 283; "Canadian Production of Petroleum Products and Natural Gas, 1913," *Intercolonial Gas Journal*, Apr 1914, 149; and F.D. Adams, "Waste of Natural Gas," *Intercolonial Gas Journal*, Apr 1915, 132.

25 "President V.S. McIntyre's Retiring Address," *Intercolonial Gas Journal*, Sept 1920, 335-6; G.W. Allen, "Possibilities Ahead for the Gas Industry," *Intercolonial Gas Journal*, Sept 1918, 337-40; and "Thirteenth Annual Convention a Real Success," *Intercolonial Gas Journal*, Sept 1920, 333-4.

Chapter Three

1 On the Canadian economy and society from 1918 to 1929, see the relevant yearly volumes of the *Canada Yearbook*, along with the history and survey of postwar reconstruction published in the 1920 and 1943 editions. Also see Brown and Cook, *A Nation Transformed*, 294-338; Norrie and Owram, *A History of the Canadian Economy*, 441-74; and J.L. Finlay and D.N. Sprague, *The Structure of Canadian History*, 6th ed. (Scarborough: Prentice-Hall, 2000), 363-76.

2 *Canada Yearbook* (1921), 338, 374, and 385; *Canada Yearbook* (1929), 353; *Canada Yearbook* (1930), 406; and *Canada Yearbook* (1931), 383-4.

3 Natural Gas Act, *Statutes of Ontario*, 9 George V, Chapter 12 (Toronto: King's Printer, 1919); NGPA, "A Review of the Natural Gas Situation in Western Ontario," *Intercolonial Gas Journal* (hereinafter, *ICGJ*), 1 Jan 1922, 24-5; and Lauriston, *The Blue Flame of Service*, 62.

4 On the background and results of the Ontario election of 1919, see Randall White, *Ontario, 1610-1985: A Political and Economic History* (Toronto: Dundurn Press, 1985), 209-18; and Kerry Badgley, *Ringing in the Common Love of Good: The United Farmers of Ontario, 1914-1926* (Montreal and Kingston: McGill-Queen's University Press, 2000).

5 With respect to the Drury government's natural gas policies, see NGPA, "A Review of the Natural Gas Situation," 24-6; Lauriston, *The Blue Flame of Service*, 62-4; and Miller, *Union Gas*, 18-19. In regard to gas shortages, also see "Report on Windsor, Chatham, and Sarnia Gas Supply," *ICGJ*, Dec 1919, 461.

6 On the Union-municipal confrontation, see NGPA, "A Review of the Natural Gas Situation," *ICGJ*, 24-6; Lauriston, *The Blue Flame of Service*, 63-4; and Miller, *Union Gas*, 18-23. Also, see "Will Carry Out Threat to Shut Off Gas," *Globe and Mail*, 1 November 1920; "Toronto Order Soon Restores Gas Services," *Globe and Mail*, 2 November 1920; "An Armistice in Gas Fight Now Obtained," *Globe and Mail*, 3 November 1920; and "Set Gas Price During Winter Pending Probe," *Globe and Mail*, 10 November 1920.

7 "Report on the Present and Future Conditions of Kent Natural Gas Field, Ontario, Canada," *ICGJ*, 1 Aug 1921, 308-10; and "How to Keep the Natural Gas Industry Alive," *ICGJ*, 1 Jan 1922, 20.

8 Natural Gas Act, *Statutes of Ontario*, 11 George V, Chapter 16 (Toronto: King's Printer, 1921); and Lauriston, *The Blue Flame of Service*, 64-5.

9 Lauriston, *The Blue Flame of Service*, 71-2; Miller, *Union Gas*, 7-10; and Christine Smith, "The First Sixty-Five Years," *The Pilot*, 1 July 1977, 17-18.

10 On Northwestern, see ATCO Archive, Northwestern Utilities, *History and Information Book*, A1-A6; Julian Garret, "Edmonton's Natural Gas System, Early History," *The Courier*, Apr 1938, 1-14; "In Memoriam," *The Courier*, July 1932, 1-13; "C.J. Yorath: Head of Calgary Gas Company – Many Western Light, Power Concerns Dead," *The Albertan*, 4 April 1932; and Stahl, *Record of Service*, 18-21.

11 Reading, "Oil Pathfinders at Turner Valley had 10 Difficult Years"; and McKenzie-Brown, Jaremko, and Finch, *The Great Oil Age*, 36-8.

12 "Gas Waste Equals 87$^{1/2}$ Cars of Coal Daily," *Calgary Herald*, 11 May 1928; "Canada's Greatest Oil Field Worthy of Visit by Tourists," *Calgary Herald*, 2 June 1928; Paul Reading, "Wealth, Waste and Peril All Pouring from Quiet Valley," *Calgary Herald*, 6 April 1929; and McKenzie-Brown, Jaremko, and Finch, *The Great Oil Age*, 38.

13 "Alberta Gas May Be Piped to Winnipeg," *Edmonton Journal*, 22 June 1928; "Urges Piping of Valley Gas to Vancouver," *Calgary Herald*, 4 March 1930; "Calgary

Concern to Export Gas to US If License Granted, *Calgary Herald*, 16 May 1929; "Will use Waste Gas in Alberta," *Calgary Herald*, 6 March 1929; and "Gas Export to Montana," *ICGJ*, February 1931, 66.

14 H.S. Tims, "Reminiscences of the Canadian Western Natural Gas Company," *The Courier*, Oct 1932, 1-2; and Stahl, *Record of Service*, 12-13.

15 "Too Many Oil Wells Are Sunk in Turner Valley, Coste Says," *Calgary Herald*, 6 March 1929; "Home Sweet Heritage Home," www.collections.ic.gc.ca/calgary/res16.htm, 12 April 2006; and "Cornerstones: Coste House," www.calgarypubliclibrary.com/calgary/historic_tours/corner/cos.htm, 12 April 2006.

16 S.E. Slipper, "A Quarter Century of Gas Supplies," *The Courier*, July 1936, 10-11; and Stahl, *Record of Service*, 13-14.

17 A.A. Dion, "Carrying On by Means of Improved Plant and Equipment," *ICGJ*, Sept 1920, 351-4; "Ottawa Power Grows; Survives Competition in Capital Metropolis," *ICGJ*, June 1931, 213; F.H. Hewlings, "Reasons for the Introduction of a Small Vertical Retort Plant in Victoria," *ICGJ*, Oct 1923, 383-4; "Victoria Gas Works," *ICGJ*, Oct 1923, 384-7; "Gas, The Modern Fuel: Its History in Vancouver, BC," *ICGJ*, Apr 1926, 143; John Keillor, "New BC Electric Gas Plant is Finished," *ICGJ*, Mar 1925, 97-8; "Improvements and Extensions at St. Thomas," *ICGJ*, May 1922, 169; and Quebec Power Co. Spend $2,500,000 on Improvements," *ICGJ*, Nov 1929, 430.

18 Lauriston, *The Blue Flame of Service*, 41, 69 and 74; and Government of Alberta, "Turner Valley Gas Plant," www.cd.gov.ab.ca/enjoying_alberta/museums_historic_-sites/site_listings/turner_valley/history/1927_47.asp, 15 April 2006.

19 Alex Forward, "Mankind's Greatest Country," *ICJG*, Aug 1925, 292; J. Keillor, "The Development of the Gas Appliance Business in Vancouver, BC," *ICGJ*, Nov 1925, 429; and Northwestern Utilities, *History and Information Book*, 69.

20 H. Hill, "Can One Man Start an Industrial Department?" *ICJG*, Mar 1923, 94-5; Tucker, *Consumers' Gas Company*, 93-8; "Cooking in Volume with Gas Fuel," *ICGJ*, Apr 1929, 125-8; John E. Philpott, "Selling Gas to the Hotels, Institutions, and Public Schools," *ICGJ*, Sept 1919, "Gas Plays an Important Part in Toronto Newspaper Offices," *ICGJ*, May 1926, 173.

21 "Consumers' Gas Company Starts New Home Service Department," *ICGJ*, May 1925, 173; Marcella P. Richardson, "The Ottawa Gas Company Home Service: Its Aims, Activities, and Achievements," *ICGJ*, July 1929, 257; Stahl, *Record of Service*, 15; and "Gas Industry Offers Many Opportunities for Women," *ICGJ*, Sept 1925, 337. Also, see Ada Bessie Swan, "The Value of a Home Service Department to a Gas Company," *ICGJ*, Nov 1925, 412-5; "Edmonton's New Service Department," *ICGJ*, Apr 1930, 4; and Hesperia Lee Aylsworth, "Why a Cooking School," *The Courier*, July 1931, 16.

22 Keillor, "The Development of the Gas Appliance Business," 429. On the implementation of more active sales efforts more generally, also see E.R. Hamilton, "Prize Contests as a Means of Stimulating New Business," *ICGJ*, Sept 1923, 328-9; Alton Jones, "The Value of a New Business Department," *ICGJ*, Sept 1926, 326; and A.G. Davis, "The Function of Salesmanship in the Development of Industrial Gas Business," *ICGJ*, Oct 1926, 371.

23 CGA, "Executive Committee Report, Year Ending June 21st, 1928," *ICGJ*, July 1928, 247; "Report of the Secretary Treasurer," *ICGJ*, July 1928, 248-9; and "Secretary-Treasurer's Report," *ICGJ*, Nov 1914, 411.

24 CGA, "Executive Committee Report, Year Ending June 21st, 1928," 246; "Report of the Canadian Blue Start Seal of Approval Committee," *ICGJ*, July 1928, 248; "Extensions to the AGA Laboratory," *ICGJ*, Apr 1928, 129-30; "Gas Laboratory to Be the Largest of Its Kind in the World," *ICGJ*, Sept 1928, 337; and "Report of the Chairman and Secretary, Laboratory Approval Committee," *ICGJ*, July 1929, 248-50.

25 "Sixteenth Annual Convention of the Canadian Gas Association," *ICGJ*, Sept 1923, 326; W.C. Beckjord, "Where the Gas Industry Stands in Research," *ICGJ*, Nov 1928, 405; B.F. Haanel, "The Federal Government's New Fuel Research Laboratory," *ICGJ*, Dec 1928, 447-9; "Economic Aspects of the Laboratory Seal," *ICGJ*, Dec 1928, 449-51; and Library and Archives Canada (LAC), MG28-I230, vol. 150, Canadian Manufacturers' Association (CMA), *The First Hundred Years* (Toronto, c1971).

Chapter Four

1 On the causes and consequences of the Great Depression, see the Royal Commission on Dominion-Provincial Relations, *Report*, Vol. 1 (Ottawa: King's Printer, 1940); J.H. Thompson and Allen Seager, *Canada, 1922-1939: Decades of Discord* (Toronto: McClelland and Stewart, 1985), 193-7 and 341-51; Norrie and Owram, *A History of the Canadian Economy*, 475-89; and Michael Bliss, *Northern Enterprise: Five Centuries of Canadian Business* (Toronto: McClelland and Stewart, 1987), 419-20.

2 *Canada Yearbook* (Ottawa: King's Printer, 1937), 381; "Products of the Coke and Gas Industry in Canada, 1928 and 1929," *ICGJ*, Nov 1930, 407; and "The Coke and Gas Industry in Canada, 1934," *ICGJ*, Oct 1935, 276.

3 "The Coke and Gas Industry in Canada for 1929," *ICGJ*, Nov 1930, 428; "The Coke and Gas Industry in Canada, 1934," *ICGJ*, Oct 1935, 276; "Gas Industry Continues to Show Growth," *ICGJ*, Oct 1938, 223; and "Canadian Gas Convention a Real Practical Meeting: The President's Address," *ICGJ*, July 1937, 171.

4 Lauriston, *The Blue Flame of Service*, 78-82; Miller, *Union Gas*, 9-10; "Plant's Closing Feared by Citizens," *ICGJ*, Oct 1937, 270-1; "Editor Follows Up Hydro's Notice to Close Cobourg Plant," *ICGJ*, June 1938, 132-3; and "Cobourg Lost Chance to Show What Gas Can Do," *ICGJ*, Jan 1939, 14.

5 "Official Opening of Guelph's New $120,000 Plant," *ICGJ*, Feb 1932, 55-6; "An Interesting Account of Guelph's Gas Dept.," *ICGJ*, Feb 1932, 56; "A Bit of Guelph's History as to in West's Gas," *ICGJ*, Feb 1933, 57-8; and "Guelph's Verticals Enable Gas Rate to be Progressively Reduced," *ICGJ*, Feb 1934, 49.

6 "City of Sherbrooke: Annual Report of the Gas Department," *ICGJ*, June 1939, 114; "Encouraging Improvement in Peterborough Gas Department," *ICGJ*, Mar 1938, 59; City of Kingston, "Coal Tar," www.cityofkingston.ca/residents/environments/coaltar/index.asp, 21 May 2006; Kitchener Utilities, "Our History," www.city.kitchener.on.ca/utilities/anniversary_1000th.htm, 20 May 2006; "Proceedings of the NGPA," *ICGJ*, Jan 1930, 25-28; R.L. Bevan, "Dayton Oil Gas Plant for Windsor," *ICGJ*, Apr 1930, 130-1; "The Dayton Oil Gas Process of General Oil and Gas Corporation," *ICGJ*, Mar 1931, 99-102. Also see E.W. King, "Some Possibilities in the Design of a Small Gas Plant," *ICGJ*, Sept 1931, 325-9; and Ellis Mills, "The Future of the Small Gas Company," *ICGJ*, Mar 1936, 64-5.

7 On the Montreal Light, Heat, and Power Company, the Winnipeg Electric Street Railway Company, and the BC Electric Street Railway Company, see Armstrong and Nelles, *Monopoly's Moment*, 93-107; and Jack Jedwab, "Louis-Joseph Forget," *DCB*.

8 On the Hamilton situation, see the Hamilton Heritage Project, "Hamilton Gas Light Company," www. collections.ic.gc.ca/industrial/19-29.htm; "Union Gas Co.," *ICGJ*, May 1935, 139; "Approve Plan for United Fuel," *ICGJ*, Jan 1939, 16; and Lauriston, *The Blue Flame of Service*, 83-4.

9 Geoff Milburn et al., *A Tradition of Service: The Story of Consumers' Gas* (Toronto: Consumers' Gas Company, 1993), 83; Arthur Hewitt, "Fuel Sales Rise in Unison with Business," *ICGJ*, Feb 1935, 44-5; and Arthur Hewitt, "Canadian Gas Industry, 1935," *ICGJ*, Feb 1936, 51.

10 H. McNair, "The Gas Industry in the Dominion of Canada," *ICGJ*, Oct 1933, 324-5; O.L. Maddux, "Gas Merchandising and Its Relation to Distribution Load," *ICGJ*, Nov 1931, 426-9; and O.L. Maddux, "Planning and Executing an Industrial Sales Programme," *ICGJ*, Oct 1933, 333-5.

11 "Uses of Gas in Industry Are Many and Increasing" and "Gas Fuel for Beauty Shops," *Saturday Night*, 6 May 1939, 27-8; H.E. Watson, "Building Gas Load in Pressing, Cleaning, and Dyeing Establishments," *ICGJ*, Apr 1936, 89-104; O.L. Maddux, "1933 Report of the Competitive Fuels Committee," *ICGJ*, Nov 1933, 351-2; and J. Fremont, "The Use of Manufactured Gas in Industrial Bake Ovens," *ICGJ*, Jan 1936, 3-6.

12 ATCO archive, Northwestern Utilities, *History and Information Book*, 68-78.

13 "Gas vs. Electric Ranges," *ICGJ*, July 1934, 218; "Gas Industry Forges Ahead," *ICGJ*, Feb 1936, 46; "Proceedings of the NGPA," *ICGJ*, Mar 1931, 103-6; and Jessie McQueen, "Home Service: In Step with Sales," *ICGJ*, Dec 299-305.

14 Maddux, "Gas Merchandising and Its Relation to Distribution Load"; "Water Heaters Make for Big Gas Loads," June 1939, 123; "Unable to Keep up with the Demand for Gas Water Heaters," June 1935, 164; and "Dominion Statistics Show More Electric than Gas Water Heaters," ICGJ, July 1934, 194.

15 O.L. Maddux, "Promotional Rates for Gas Services," *ICGJ*, Jan 1935, 10-14; Roy Soderlind, "House Heating with Manufactured Gas," *ICGJ*, Oct 1930, 365-70; Maddux, "1933 Report of the Competitive Fuels Committee," 352-3; and "Gas Heated Buildings: Ten Years of Progress in Vancouver," *ICGJ*, Nov 1933, 368-73.

16 On King's record and the election of 1930, see H. Blair Neatby, "William Lyon Mackenzie King," *DCB*; P.B. Waite, "Richard Bedford Bennett," *DCB*; Thompson and Seager, *Canada 1922-1939*, 197-205; and J. Murray Beck, *Pendulum of Power* (Scarborough: Prentice Hall, 1968), 191-205.

17 On public policies, government revenues, and the election of 1935, see Thompson and Seager, *Canada 1922-1939*, 253-302; *Canada Yearbook*; and Beck, *Pendulum of Power*, 206-222.

18 BC Hydro Power Pioneers, *Gaslights to Gigawatts: A Human History of BC Hydro and Its Predecessors* (Vancouver: Hurricane Press, 1998); and Milburn et al, *A Tradition of Service*, 86. Also see the *Labour Gazette*, c 1930.

19 Lauriston, *The Blue Flame of Service*, 82; Milburn et al., *A Tradition of Service*, 83; and "North Western Utilities," *The Courier*, July 1940, 34.

20 See R.C. Sparling, "Plumber Cooperation in Toronto," *ICGJ*, Mar 1931, 85-7; M.L. Kane, "Sales Promotion Work for the Future of the Gas Industry," *ICGJ*, Oct 1934, 305-8; W.C. Mainwaring, "Intensive Sales Development," *ICGJ*, Sept 1936, 225-8; J. Lightbody, "The Modern Fuel Needs Modern Advertising," *ICGJ*, Dec 1936, 296-99; and Frank D. Howell, "The Electric Bogey," *ICGJ*, 135-9.

21 *Canada Yearbook* (1941), 248, 265, and 322.

22 Julian Garret, "The Natural Gas Industry in Alberta," *ICGJ*, Sept 1939, 218; R.B. Harkness, "Natural Gas in Quebec," *ICGJ*, Oct 1931, 365-6; "Quebec Surveys St. Lawrence Oil and Gas Fields," *ICGJ*, Oct 1931, 383-4; and "Natural Gas in the St. Lawrence Valley," *ICGJ*, June 1933, 199-200.

23 On Turner Valley, see "Waste Gas Problems," *ICGJ*, Oct 1930, 386; "Huge Waste of Gas in Alberta," *ICGJ*, Sept 1937, 236; P.D. Mellon, "Storage of Natural Gas in the Bow Island Field," *ICGJ*, 275-8; "Will Limit Output to Conserve Alberta Gas," *ICGJ*, Nov 1938, 258. Also, see Supreme Court of Canada, *Reports* (Ottawa: King's Printer, 1933); and John Richards and Larry Pratt, *Prairie Capitalism: Power and Influence in the New West* (Toronto: McClelland and Stewart, 1979), 54-8.

24 "Notwithstanding Distance, Canadian Meeting a Success," *ICGJ*, July 1930, 255-6; "Proceedings of the Twenty-Ninth Annual Convention of the Canadian Gas Association and Northwest Conference of Pacific Coast Gas Association at Vancouver," *ICGJ*, Aug 1936, 195-200; T.P. Pinckard, "President's Address," ICGJ, July 1939, 133; and "Congratulations to Mr. Arthur Hewitt," *ICGJ*, Dec 1934, 388.

25 "New Constitution and Bylaws of the Canadian Gas Association Opens Up a Wide Field of Usefulness to the Whole Gas Industry," *ICGJ*, Dec 1935, 316-9; "Canadian Gas Adopts New Constitution," Feb 1936, 49; and "Almost Every Gas Utility in Canada Approved New Constitution," *ICGJ*, Feb 1936, 50.

Chapter Five

1 See *Canada Yearbook* (1935-9 and 1945); and Edward J. Tucker, "The Accomplishments of the Canadian Gas Industry in World War II," *Canadian Gas Journal (*hereinafter, CGJ), Aug 1946, 159-64.

2 On Canada's war effort, see *Canada Yearbook* (1946); C.P. Stacey, *Arms, Men, and Governments: The War Policies of Canada, 1939-1940* (Ottawa: The Queen's Printer, 1970); *Canada and the Second World War: Valour Remembered, 1939-1945* (Ottawa: Ministry of Supply and Services, 1981); Robert Bothwell, Ian Drummond, and John English, *Canada, 1900-1945*, 349-87; and Ruth Roach Pierson, *They're Still Women After All: The Second World War and Canadian Womanhood* (Toronto: McClelland and Stewart, 1986).

3 See Leacy, Buckley, and Urquhart, *Historical Statistics of Canada*, H19-34; and Peter S. McInnis, *Harnessing Labour Confrontation: Shaping the Postwar Settlement in Canada, 1943-1950* (Toronto: University of Toronto Press, 2002), 24.

4 *Canada Yearbook* (1943-5); Bothwell, Drummond, and English, *Canada, 1900-1945*, 352; McInnis, *Harnessing Labour Confrontation*, 24; and Pierson, *They're Still Women After All*, 28-9 and 69-71.

5 McInnis, *Harnessing Labour Confrontation*, 23-5.

6 G.W. Allen, "The Canadian Gas Industry Improves Its Position and Plays Important Part in Wartime Heat Requirements, *CGJ*, Nov 1941, 257-8; and T.P. Pinckard, "Activities of the Gas Industry in Wartime," *CGJ*, July 1942, 148-50.

7 Allen, "The Canadian Gas Industry Improves" 257.

8 *Ibid.*; and CGA Executive Committee, Minutes, 15 June 1943, CGA archive, Vol. 2, 45.

9 Edward J. Tucker, "Our Job in the Present Emergency," *CGJ*, June 1941, 126-9; CGA Executive Committee, *Minutes*, 8 December 1942, CGA archive, Vol. 2, 34. Also, see "Industrial Nutrition," *CGJ*, Oct 1942, 217-8; "Servel Begins Advertising Nutrition in Industry," *CGJ*, Dec 1942, 261-2; and George S. Jones, "Our Part in the War Programme," *CGJ*, Mar 1943, 48.

10 Julian Garret, "The Northwestern Utility Gas Company's Contribution to the War Effort," *CGJ*, July 1943, 123-5; and Edmonton Public Library, "Population of Edmonton, 1878-2002," www.epl.ca/Elections/ info.EPL.population.cfm, 26 June 2005.

11 "War Boosts Gas Business," *CGJ*, July 1941, 151-3; and "Gas Industry Meets Big Changes," *CGJ*, Apr 1943, 63-5.

12 Bothwell, Drummond, and English, *Canada, 1900-1945*, 357-60; "War Boosts Gas Business"; Tucker, "Our Job in the Present Emergency,"; and "War Brings New Uses for Gas," *CGJ*, Jan 1943, 5-6.

13 On gas workers, Price, and Canadian casualties, see Edward J. Tucker, "The Accomplishments of the Gas Industry in World War II," *CGJ*, Aug 1946, 159-64; "P.O. Jack Price Gets Back to England in Damaged Plane," *The Courier*, June 1943, 29; Scott Robertson, "In the Shadow of Death by Moonlight," in David J. Bercuson and S.F. Wise (eds.), *The Valour and the Horror Revisited* (Montreal and Kingston: McGill-Queen's University Press, 1994), 163; and *Canada Yearbook* (1945).

14 Tucker, "The Accomplishments of the Gas Industry,"; Pinckard, "Activities of the Gas Industry in Wartime,"; "What about Taxation?" CGJ, Sept 1941, 206.

15 On gas industry statistics in the Second World War, see "War Boosts Gas Business"; "Gas Industry Meets Big Changes"; *Canada Yearbook* (1945), 529; and Tucker, "The Accomplishments of the Gas Industry."

16 William Lyon Mackenzie King, *Canada and the War: Victory, Reconstruction, and Peace* (Ottawa, 1945), 88-9.

17 Leacey, Buckley, and Urquhart, *Historical Statistics of Canada*, Y211-259.

18 See William Beveridge, *Social Insurance and Allied Services in the United Kingdom* (London: King's Printer, 1942); and Leonard Marsh, *Report on Social Security for Canada* (Ottawa: King's Printer, 1943).

19 On postwar planning and reconstruction, see Doug Owram, *The Government Generation: Canadian Intellectuals and the State* (Toronto: University of Toronto Press, 1986); Greg Donaghy (ed.), *Uncertain Horizons: Canadians and Their World in 1945* (Ottawa: Canadian Committee for the History of the Second World War, 1997); and Tim Krywulak, "An Archaeology of Keynesianism: The Macro-Political Foundations of the Modern Welfare State" (PhD diss., Carleton University, 2005).

20 "Investing in Utilities: Postwar Developments Will Be a Major Factor," *CGJ*, Jan 1943, 9-10; C.V. Sorenson, "General Resume of Postwar Planning Committee of the American Gas Association," *CGJ*, July 1943, 132-5; "Postwar Planning," *CGJ*, Nov 1943, 204-8; and "Postwar Committee Meeting of the Canadian Gas Association," *CGJ*, Dec 1943, 230.

21 Executive Committee, Minutes, 11 Oct 1939 and 2 July 1940, CGA archive, Vol. 2, 5 and 15; "War Brings New Uses for Gas"; "President Garret's Address at the CGA

Convention at Jasper Park Lodge, Alberta," *CGJ*, Aug 1940, 182-6; Tucker, "The Accomplishments of the Gas Industry"; and G.W. Allen, "Gas Industries Are Preparing Postwar Plans," *CGJ*, Jan 1945, 3-5.

22 "What Are Women Thinking," *CGJ*, Mar 1941, 64-7; and "Flame for Sale," *CGJ*, Nov 1946, 259. Also see, for instance, Walter C. Beckjord, "Development of a Sound Sales Policy," *CGJ*, July 1940, 155-8; "How to Win the Sweepstakes," *CGJ*, Mar 1941, 51-6; R.E. Williams, "Battles Are Won by Well Trained Troops," *CGJ*, Jan 1945, 13-15; and "Miss Jessie McQueen Discusses Paper on Planning for Postwar Sales," *CGJ*, Aug 1945, 151-4.

23 "Men and Machines Keep Gas Production Going at Winnipeg Electric Gas Works," *CGJ*, May 1949, 114-5; Alan H. Harris, "Winnipeg Electric Company's New Propane Gas Plant," *CGJ*, May 1946, 87-8; "New Propane Plant at Victoria," *CGJ*, May 1949, 125; "Gas Sales in Canada in 1947 Reach an All-Time Peak," *CGJ*, June 1948, 151; G.W. Allen, "Gas Industry Forging Ahead and Public Appreciation Still Strong," *CGJ*, Dec 1948, 291-3; and *Canada Yearbook* (1950).

24 NGPA, Proceedings (1944), Union Gas archive; Lauriston, *The Blue Flame of Service*, 89-102; and "Importing 400 Million Cubic Feet a Month From Texas," *CGJ*, Aug 1950, 228-9.

25 H.R. Milner, "The Natural Gas Resources of Alberta," *CGJ*, Aug 1948, 195; "Progress in the Development of Long Distance Gas Transmission Lines," *CGJ*, July 1931, 257-61; Lauriston, *The Blue Flame of Service*, 89-102; and Peter C. Newman, *The Promise of the Pipeline* (Calgary: TransCanada Pipelines, 1993), 5-8.

26 Executive Committee, Minutes, 1939-1949; and "President Garret's Address at the CGA Convention," *CGJ*, Aug 1940, 182-6.

Chapter Six

1 On the construction of the TransCanada pipeline, see Newman, *The Promise of the Pipeline*, i-ii and 19-23; and Kilbourn, *Pipeline*, 159-64.

2 Province of Alberta, Natural Gas Commission: *Enquiry into Reserves and Consumption of Natural Gas in Alberta* (Edmonton: King's Printer, 1949), 12-17. Also, see Newman, *The Promise of the Pipeline*, 7-8; and Kilbourn, *Pipeline*, 17-19.

3 Province of Alberta, *Statues of Alberta* (Edmonton: King's Printer, 1949); Vince Fitzpatrick, *BC Gas: Fifty Years of Commitment* (Vancouver: BC Gas, 2002), 4-5; and Kilbourn, *Pipeline*, 17, 20, 26, and 34-44.

4 On the creation of TransCanada Pipelines, see Robert Bothwell and William Kilbourn, *C.D. Howe: A Biography* (Toronto: McClelland and Stewart, 1979), 283-4; and Kilbourn, *Pipeline*, 25-44.

5 P.B. Waite, *Canada: 1874-1896: Arduous Destiny* (Toronto: McClelland and Stewart, c 1971); Norrie and Owram, *A History of the Canadian Economy*, 224-32 and 299-311; and William Marr and Donald Paterson, *Canada: An Economic History* (Toronto: Gage, 1980), 317-31.

6 Newman, *Promise of the Pipeline*, 10-14; and Kilbourn, *Pipeline*, 65-116.

7 House of Commons, *Debates*, 7 May-7 June 1956 (Ottawa: Queen's Printer, 1957); Newman, *Promise of the Pipeline*, 13-18; Bothwell and Kilbourn, *C.D. Howe*, 299-316; Kilbourn, *Pipeline*, 111-33; and Finlay and Sprague, *The Structure of Canadian History*, 449.

8 Bothwell and Kilbourn, *C.D. Howe*, 317-31; Kilbourn, *Pipeline*, 152-66; Finlay and Sprague, *The Structure of Canadian History*, 443-52; Denis Smith, *Rogue Tory: The Life and Legend of John. G. Diefenbaker* (Toronto: Macfarlane Walter and Ross, 1995), 188-241; John Meisel, *The Canadian General Election of 1957* (Toronto: University of Toronto Press, 1957).

9 Kilbourn, *Pipeline*, 159; "Union Signs Contract," *CGJ*, Mar 1955, 17; Miller, *Union Gas*, 101-3; Milbourn, et al., *Tradition of Service*; Hydro-Québec Archives, "Fonds Commission Hydroélectric Québec, 1944-1963," site.rdaq.qc.ca/ArchivesHydroQuebec, 4 August 2006; and "Québec Natural Gas Conversion Program Underway," CGJ, Feb 1958, 10-11.

10 Charles Hay, "Postwar Development of the Natural Gas Industry," in Hilborn, *Dusters and Gushers*, 193; and DBS, *Sales of Manufactured and Natural Gas*, Catalogue No. 45-005 (Ottawa, December 1963).

11 National Energy Board Act, *Statutes of Canada*, (1959); Ralph Toombs, "Canadian Energy Chronology," Natural Resources Canada, www2.nrcan.gc.ca/es/es/ EnergyChronology/index_e.cfm, 15 August 2006; and National Energy Board, *Twenty-Five Years in the Public Interest* (Ottawa: Department of Natural Resources, 1982).

12 Otto Zwanzig, "Gas Users Served Best by No Rate Regulation," *CGJ*, Nov 1960, 28-31 and 60; and "Consumer's Welfare Is Main Case for Gas Regulation," *CGJ*, Dec 1960, 22-3.

13 Alastair Lucas and Trevor Bell, *The National Energy Board: Policy, Procedure, and Practice* (Toronto: Law Reform Commission of Canada, 1977). Also see "lists of submissions" for the Royal Commission on Canada's Economic Prospects and the Royal Commission on Energy.

14 On membership numbers, see Executive Committee, *Minutes*, 27 Nov 1953, 3.

15 D.K. Yorath, "CGA Must Speak with Own Voice," *CGJ*, Aug 1954, 17.

16 *Ibid.*, 7 May-20 June 1954; "Over 500 People Attend CGA Annual Convention," *CGJ*, Aug 1954, 12.

17 See, *Constitution and Bylaws of the Canadian Gas Association* (CGA: Toronto, 1969) [Includes all revisions from 1956-1969].

18 D.K. Yorath, 13-15; "Leaders Look to the Future as Canadian Gas Association 48th Annual Convention Opens," *CGJ*, July 1955, 7-9; and Executive Committee, Minutes, 22 Aug 1955, 3.

19 Executive Committee, Minutes, 19 June 1950, 20 April 1951, and 18 June 1951; and CGA "Success of the Gas Industry ... Rests Entirely Upon How Gas Companies Support the Plan to Sell Nothing but Approved Appliances," *CGJ*, Mar 1952, 52.

20 Executive Committee, Minutes, 8 Oct 1954-13 Jan 1956.

21 *Ibid.*, 16 Apr 1956-4 Apr 1957; "CGA Meets Provincial Representatives in Quest for Gas Standards Approval," *CGJ*, June 1956, 23; "BC, Ontario Adapting CGA Approvals Program," *CGJ*, Sept 1956; and "The Canadian Gas Association's Approvals Program, Report No. 1," *CGJ*, Mar 1957, 10-13.

22 Executive Committee, *Minutes*, 16 Apr 1956-5 Apr 1957; "Gas Technology Course Passes Initial Run," *CGJ*, Apr 1960, 16-18; and "Gas Technology Grads Honored," *CGJ*, May 1960, 33.

23 Executive Committee, Minutes, 24 June 1954-15 Feb 1955 and 13 Jan 1956-17 Sept 1956.

24 *Ibid.*, 24 June 1957.

Chapter Seven

1 Norrie and Owram, *A History of the Canadian Economy*, 549.

2 John Davis, *Canadian Energy Prospects* (Ottawa: Royal Commission on Canada's Economic Prospects, 1957), 1-12 and 239-64; and Robert Bothwell, *Nucleus: The History of Atomic Energy of Canada Limited* (Toronto: University of Toronto Press, 1988), 182-211.

3 With respect to the history of nuclear development from late the 1940s to the early 1980s, see Bothwell, *Nucleus*, 54-451; Ronald Babin, *The Nuclear Power Game* (Montreal: Black Rose Books, 1985), 43-90; Gord L. Brooks, "A Short History of the CANDU Nuclear Power System," Canteach, canteach.candu.org, 7 Sept 2006; and Alan Nixon, "The Canadian Nuclear Power Industry," Government of Canada, Ministry of Industry, Science, and Technology, BP-365e, dsp-psd.pwgsc.gc.ca/Collection-R/LoPBdP/BP/bp365-e.htm, 9 Sept 2006.

4 Davis, *Canadian Energy Prospects*, 239-64; Bothwell, *Nucleus*, 208-11; Babin, *The Nuclear Power Game*, 48-90; and Nixon, "The Canadian Nuclear Power Industry," A-E.

5 Davis, *Canadian Energy Prospects*, 8; and CGA, *Canadian Gas Facts* (1981), 3.

6 Toombs, "Canadian Energy Chronology," www2.nrcan.gc.ca/es/es/EnergyChronology/index_e.cfm, 15 August 2006. Also see the 1961-9 editions of DBS, *Canada Yearbook*.

7 J.W. Kerr, "Keeping up the Pressure ... From Now Until 1970," CGA archive, annual convention files; and CGA, *Canadian Gas Facts* (1981), 12.

8 For Canadian gas statistics, see the CGA's *Canadian Gas Facts* (1981-4). On Canada-US relations, see G. Norman Hillmer (ed.), *Partners Nevertheless: Canadian-American Relations in the Twentieth Century* (Toronto: Copp Clark Pitman, 1989); Charles F. Doran and John H. Sigler (eds.), *Canada and the United States: Enduring Friendship, Persistent Stress* (New Jersey: Prentice Hall, 1985); and Robert Bothwell, *Canada and the United States: The Politics of Partnership* (Toronto: University of Toronto Press, 1992).

9 G. Norman Hillmer and Maureen Appel Molot (eds.), *Canada Among Nations, 2002: A Fading Power* (Don Mills: Oxford University Press, 2002), 1-33; and Andrew Cohen, *While Canada Slept: How We Lost Our Place in the World* (Toronto: McClelland and Stewart, 2003).

10 See Dennis Guest, *The Emergence of Social Security in Canada*, 2nd rev. ed. (Vancouver: University of British Columbia Press, 1985); and Janine Brodie, *The Political Economy of Canadian Regionalism* (Toronto: Harcourt Brace Jovanovich, 1990).

11 Hillmer and Molot, *Canada Among Nations*, 1-33; and Department of Foreign Affairs and International Trade, "Peacekeeping Missions over the Years and Canada's Contribution," www.dfait-maeci.gc.ca/peace keeping/missions-en.asp, 12 Sept 2006.

12 Buckley, Urquhart, and Leacy (eds.), *Historical Statistics of Canada*, H19-34; and Andrew Coyne, "Social Spending, Taxes, and the Debt: Trudeau's Just Society" and Linda McQuaig, "Maverick Without a Cause: Trudeau and Taxes," in Andrew Cohen and J.L. Granatstein (eds.), *Trudeau's Shadow: The Life and Legacy of Pierre Elliott Trudeau* (Toronto: Random House of Canada, 1998), 223-55.

13 "2nd General Session, 56th Annual Meeting, 25 June 1963," CGA archive, annual conventions files.

14 Buckley, Urquhart, and Leacy (eds.), *Historical Statistics of Canada*, G188-202 and G415-428.

15 Richard G. Lipsey, "Canada and the United States: The Economic Dimension," in Doran and Sigler (eds.), *Canada and the United States*, 69-108; Robin Mathews and James Steele, *The Struggle for Canadian Universities* (Toronto: New Press, 1969); and Jeffery Simpson, *Star-Spangled Canadians: Canadians Living the American Dream* (Toronto: HarperCollins, 2000), 1-34.

16 On foreign investment, see the Royal Commission on Canada's Economic Prospects, *Report*; Melville Watkins et al., *Foreign Ownership and the Structure of the Canadian Industry* (Ottawa: Privy Council Office, 1968); and Herbert Gray, *Foreign Direct Investment in Canada* (Ottawa: Information Canada, 1972). For examples of academic studies, see Kari Levitt, *Silent Surrender: The Multinational Corporation in Canada* (Toronto: Macmillan, 1970); and Patricia Marchak, *In Whose Interests: An Essay on Multinational Corporations in the Canadian Context* (Toronto: McClelland and Stewart, 1979).

17 Michael Howlett and M. Ramesh, *Canadian Political Economy: An Introduction* (Toronto: McClelland and Stewart, 1992), 239; and Jock Finlayson, "Canadian International Economic Policy," in Denis Stairs and Gilbert R. Winham (eds.), *Canada in the International Political/Economic Environment* (Toronto: University of Toronto Press and Ministry of Supply and Services, 1985), 51.

18 See the Mackenzie Valley Pipeline Inquiry, *Northern Frontier, Northern Homeland* (Ottawa: Ministry of Supply and Services, 1977); Peter H. Pearse (ed.), *The Mackenzie Valley Pipeline: Arctic Gas and Canadian Energy Policy* (Toronto: McClelland and Stewart, 1974); and McKenzie-Brown, Jaremko, and Finch, *The Great Oil Age*, 148-9.

19 Edgar O'Ballance, *No Victor, No Vanquished: The Yom Kippur War* (London: Barrie and Jenkins, 1979); and CBC, "The Price of Oil: In Context," www.cbc.ca/news/backgroud/oil/, 18 Sept 2006.

20 G. Bruce Doern and Glen Toner, *The Politics of Energy: The Development and Implementation of the NEP* (Toronto: Methuen, 1985), 102; Thomas Baumgartner and Atle Midttun (eds.), *The Politics of Energy Forecasting: A Comparative Study of Western Europe and North America* (Oxford: Claredon Press, 1987), 3-9; CBC, "The Price of Oil," 18 Sept 2006; and Edgar L. Jackson and L.T. Foster (eds.), *Energy Attitudes and Policies* (Victoria: University of Victoria, Department of Geography, 1980), 13.

21 Toombs, "Canadian Energy Chronology," 1971-1973; and Doern and Toner, *The Politics of Energy*, 99, 138-41.

22 Toombs, "Canadian Energy Chronology," 1971-4; and Doern and Toner, *The Politics of Energy*, 329.

23 Report of the Board of Directors (1975), CGA archive, board of directors files.

24 *Statutes of Canada* (Ottawa, 1981); Toombs, "Canadian Energy Chronology," 1981-1; Robert Bothwell, Ian Drummond, and John English, *Canada Since 1945* (Toronto: University of Toronto Press, 1982), 451-5; Doern and Toner, *The Politics of Energy*, 1-5, 105-6, and 341; and Bruce Doern and Glen Toner, "The Two Energy Crises and Oil and Gas Interest Groups: A Re-examination of Berry's Propositions," *Canadian Journal of Political Science* (1986), 481-2.

25 CGA, "The National Energy Program: Its Impact on Canadian Consumers and the Natural Gas Industry," CGA archive, policy statements files.

26 Forbes, "Crude Oil Prices, 1865-2005," www.forbes.com/static_html/oil/2004/oil.shtml, 22 Sept 2006.

27 CGA archive, board of directors files, 14 June 1981; and interview with Shahrzad Rahbar, 15 Sept 2006.

28 CGA archive, board of directors files, 23-37 June 1963.

29 CGA archive, Conference files, 26 June 1964; and board of directors files, 26-30 June 1965 and 25-28 May 1975.

30 CGA archive, Conference files, 26 June 1964; and board of directors files, 30 May 1979.

31 CGA archive, board of directors files, 23-27 June 1963.

32 *Ibid.*; and board of directors files, 26-30 June 1965 and 12-16 June 1977.

33 CGA archive, board of directors files, 23-27 June 1963 and 26-30 June 1965.

34 CGA archive, board of directors files, 12-16 June 1977.

35 CGA archive, board of directors files, 25-28 May 1975.

36 CGA archive, board of directors files, 23-27 June 1963 and 26-30 June 1965.

Chapter Eight

1 Department of Finance, *Economic Development for Canada in the 1980s* (Ottawa: Ministry of Supply and Services, 1981); and Michael Howlett and M. Ramesh, *The Political Economy of Canada*, 247-50.

2 Royal Commission on the Economic Union and Development Prospects for Canada, *Report*, Vol. 2 (Ottawa: Ministry of Supply and Services, 1985), 184-201.

3 Department of Finance, *Economic Development for Canada in the 1980s*, 1; and Department of Finance, *A New Direction for Canada: An Agenda for Economic Renewal* (Ottawa: Ministry of Supply and Services, 1984), 23.

4 Government of Canada, "The Western Accord: An Agreement Between the Governments of Canada, Alberta, Saskatchewan, and British Columbia" (Ottawa, 1985); and John F. Helliwell, Mary E. MacGregor, Robert N. McRae, and André Plourde, "The Western Accord and Lower Energy Prices," *Canadian Public Policy* 12, 2 (1986), 341-55.

5 Government of Canada, "Agreement [among the Governments of Canada, Alberta, Saskatchewan, and British Columbia] on Natural Gas Markets" (Ottawa,

1985); and Holly Reid, *Deregulation of the Canadian Natural Gas Market* (Ottawa: Public Interest Advocacy Centre, 1999).

6 Reid, *Deregulation*; and Roland Priddle, interview, 28 Nov 2006.

7 British Petroleum (BP), *BP Statistical Review of World Energy* (London, 2005).

8 Reid, *Deregulation*; and Priddle, interview, 28 Nov 2006.

9 Interviews with Priddle, 28 Nov 2006; Rudy Reidl, 15 Nov 2006; Bryan Gormley, 27 Oct 2006; and Robert Joshi, 27 October 2006. On third-party marketers and the trend toward deregulation, also see Reid, *Deregulation*; National Energy Board, *Natural Gas Market Assessment: Canadian Natural Gas, Ten Years After Deregulation* (Ottawa, 1996), available www.neb-one.gc.ca/energy/EnergyReports/EMAGas10Years-AfterDeregulation1996_e.pdf, 28 Nov 2006.

10 CAPP, *Natural Gas Prices in the North American Market* (Ottawa, 2004), available www.capp.ca/raw. asp?x=1&dt=NTV&e=PDF&dn=81664, 1 Dec 2006; NEB, *Natural Gas Market Assessment*; and Canadian Petroleum Foundation (CPF), *Canada's Pipelines* (Calgary, 2000).

11 Reidl, interview, 15 Nov 2006.

12 Interviews with Gormley, Joshi, and Michael Cleland, 27 Oct 2006.

13 Natural Resources Canada, "Frequently Asked Questions About Natural Gas Prices," www2.nrcan.gc.ca/es/erb/prb/english/ View.asp?x=448, 30 Oct 2006; Energy Shop, "Deregulation information," www.energyshop.com/es/faq/dereg.cfm, 30 Oct 2006; and CBC, "The Unnatural Price of Natural Gas," www.cbc.ca/news/background/ energy/natural_gas.html, 3 Jan 2006.

14 Reid, *Deregulation*; and Energyshop.com, "About Deregulation," www.energyshop.com/es/faq/dereg. cfm, 29 Oct 2006.

15 Energy Information Administration (EIA), "International Energy Outlook 2006," www.eia.doe.gov/oiaf/ ieo/world.html, 4 Dec 2006.

16 United Nations Environment Program, "From Rio to Johannesburg," www.uneptie.org/pc/agri-food/ WSSD/milestone.htm, 5 Dec 2006.

17 *Ibid.*; World Health Organization, "Director General, Dr. Gro Harlem Brundtland," www.who.int/ dg/brundtland/bruntland/en/index.html, 5 Dec 2006; and World Commission on Economic Development, *Our Common Future* (Oxford: Oxford University Press, 1987), available: Federal Office for Spatial Development (Switzerland), "Brundtland Report," www.are.admin.ch/are/en/nachhaltig/international_uno/unterseite02330/index.html, 5 Dec 2006.

18 Interview, Cleland, 27 Oct 2006; and Environment Canada, *Report on Plans and Priorities* (Ottawa, 1999 and 2005), available: www.ec.gc.ca/introec/dept_plng.htm, 5 Dec 2005.

19 Mike Cleland, Address to the House of Commons Environment Committee, Ottawa, 21 Nov 2006; and CGA, *'Smart Gas' Proposals for the Government of Canada* (Ottawa, 2006).

20 Shahrzad Rahbar, interview, 15 Sept 2006.

21 *Ibid.*; Cleland, interview, 27 Oct 2006.

Chapter Nine

1 On gas reserves, see Natural Resources Canada, *Canadian Natural Gas: Review of 2004 and Outlook to 2020* (Ottawa, 2006); and BP, *BP Statistical Review of World Energy* (2005).

Chairmen
1907-2006

 1907, Powell

 1908, Hay

 1909, Norris

 1910-12, Hewitt

 1913, Dion

 1914, Mann

 1915, Wallace

 1916, King

 1917, Young

 1918, Folger

 1919, McIntyre

 1920-21, Bagg

 1922, Street

 1923, Jefferis

 1924, Hamilton

 1925, Humphreys

 1926, Armstrong

 1927, Byrnes

 1928, Elcock

 1929, Dawson

 1930, Leavitt

 1931-32, McNair

 1933, Munroe

 1934, Dawson

 1935, Keillor

 1936, Munroe

 1937, Tucker

 1938, Pinckard

 1939, Garret

 1940, McNary

 1941, Pead

 1942-43, Howell

 1944, Harris

 1945, VonMaur

 1946, Weir

 1947, Brownie

 1948, MacKenzie

 1949, Sieger

 1950, Smith

 1951, Perkins

 1952, Latreille

 1953, Yorath

 1954, Severson

1955, Geldard 1956, Palin 1957, Darroch 1958, Purdy 1959, Tanner 1960, McPherson 1961, Jones 1962, Ostler 1963, Cass-Beggs

1964, Kerr 1965, Wilson 1966, Dutton 1967, Robbins 1968, McMahon 1969, Bovey 1970, McCarthy 1971, Maybin 1972, McMurrich

1973, Horte 1974, Kidd 1975, Rasmussen 1976, Capewell 1977, Dudley 1978, King 1979, Cameron 1980, Kadlec 1981, Leroux

1982, MacLeod 1983, Martin 1984, Thompson 1985, Didur 1986-87, Walker 1988, Dafoe 1989, Park 1990, Bellringer 1991-92, Farwell

1993-94, Williams 1994-96, Caillé 1996-98, Munkley 1999, Parsons 2000, Lloyd 2001, Reid 2002-03, Tessier 2004-05, Letwin 2006, Engler

Index

Page numbers followed by c *reference a photo caption; an* s *indicates a sidebar, and* t *indicates a table.*